Analysis and Compensation of Kinetic Friction in Robotic and Mechatronic Control Systems

Analysis and Compensation of Kinetic Friction in Robotic and Mechatronic Control Systems comprehensively covers the theory behind kinetic friction, as well as compensation methods and practical solutions, and serves as a key companion to studying different control systems.

Beginning with a clear introduction to the subject, the book goes on to include three main facets of kinetic friction, starting with the phenomena of kinetic friction in drives. From this, the text examines motion dynamics with friction, which introduces dynamic system equations and focuses on both energy balance and dissipation. Finally, the book explains compensation of friction in motion control, which summarises key compensation methods in controlled mechanical systems. Introducing various basic feedback control methods, including observer-based methods to compensate for kinetic friction, it provides practical information which can be used in a wide variety of contexts not specific to particular systems or applications.

The book will be of interest to students and industry workers in the fields of robotics, mechanical systems and control engineering.

Analysis and Compensation of Kinetic Friction in Robotic and Mechatronic Control Systems

Michael Ruderman

CRC Press
Taylor & Francis Group
Boca Raton London New York

CRC Press is an imprint of the
Taylor & Francis Group, an **informa** business

First edition published 2024
by CRC Press
2385 NW Executive Center Drive, Suite 320, Boca Raton FL 33431

and by CRC Press
4 Park Square, Milton Park, Abingdon, Oxon, OX14 4RN

CRC Press is an imprint of Taylor & Francis Group, LLC

ISBN: 978-1-032-53945-4 (hbk)
ISBN: 978-1-032-54070-2 (pbk)
ISBN: 978-1-003-41501-5 (ebk)

DOI: 10.1201/9781003415015

Typeset in Nimbus font
by KnowledgeWorks Global Ltd.

Contents

Preface

The idea to write this manuscript arose after being invited as a guest lecturer for the master's course, Modeling and Control of Robots, at the Department of Engineering Cybernetics of Norwegian University of the Science and Technology (NTNU) in the Spring Semester 2021. Later, the content was largely expanded and streamlined for making it into a course module aimed for mixed audiences of the PhD students and advanced master's students. The lecture materials were used four times at different institutions, within the compact and focused course modules, in 2022 and Spring 2023. The present text is a complete companion manuscript to the given lectures, yet with a significant extension of those materials related to the friction compensation. The course materials comprise approximately ten academic hours of lectures. It can be used for both a self-study and as a reading accompaniment, for example, when attending special courses in a master's program or regular courses in a doctoral program, and for all cases in robotics, mechatronics, or control engineering. Moreover, some educators or practicing engineers in the field may find the topics and content useful in one way or another, for example, for using it in their own courses or R&D projects. The prerequisites for following this text are not large scale. The reader should be familiar, at long last, with dynamical systems in general and have the background in a standard, some what classical, control theory at the bachelor's and master's levels. Furthermore, the reader have basic understanding of the actuators and drives used in robotics and mechatronics. Some prior knowledge in theoretical mechanics, in particular, contact mechanics and friction, might be advantageous, but is not required.

Materials for this course module are organized chronologically and progressively into three main parts, which follow after the brief Introduction. The latter, as the first chapter, includes several historical aspects of studying friction and also the notes about further reading on the subject. The first main part, the second chapter, Phenomena of Kinetic Friction in Drives, starts with an explanation of the friction interfaces and occurrence of the friction effects. It introduces the basic assumptions, laws, and statements for describing and modeling kinetic friction. The focus is on the classical nonlinear Coulomb and linear viscous friction at a steady-state motion, while the transient friction processes are also treated on the edge for the sake of completeness. The second main part, which is the third chapter, Motion Dynamics with Friction, considers the simple one-degree-of-freedom (1DOF) motion systems where dynamics is affected by kinetic friction. Energy balance and energy dissipation, due to friction forces, are discussed together with the transient behavior of dynamic feedback systems. The motion onset and motion stop are discussed as a matter of the damping friction forces. Attention is also paid to the convergence of state trajectories and to invariant sets, owing to the Coulomb friction of the feedback controlled motion. Finally, the third main part, the fourth chapter, Compensation of Friction in Motion Control, provides a consolidated entry into the compensation methods aimed at mitigating the parasitic side effects caused by the kinetic friction in the controlled mechanical systems. The compensation methods discussed do not require a complete redesign of a basic motion controller, and thus can be seen as a plug-in extension to either standard output- or state-feedback control. While all the given materials attempt to orient themselves also towards some recent developments, which can be found in the referred published studies, it is worth noting that the presented modeling and control approaches are purposefully chosen to be rather conservative. Only those methods and strategies that are more general and do not require very sophisticated models or compensation algorithms are included. Care is also taken to ensure a certain generality of the presented solutions, omitting those system- and application-specific features which might be cumbersome from the reader's point of view, both for education and implementation. The author would like to let the readers decide whether the proposed level of detail is sufficient for their specific friction-related problems, or whether additional literature and knowledge sources are required.

By the end of preface, I would like to express my gratitude to CRC Press (Taylor & Francis Group) for showing interest and helping to publish this manuscript, which definitely does not fit into the scope of a journal article, but at the same time is more compact and modest than a textbook for some standard one-semester courses. I am grateful to Anton Shiriaev and Roberto Oboe each one individually, for providing excellent opportunities to teach this course module and its parts to the diverse and heterogeneous audiences of students. All feedback and discussions following lectures of the material, for which I am very grateful, helped shaping it further toward a self-contained manuscript. I am also thankful to Leonid Fridman, who hosted me at UNAM, Mexico in Spring 2022, and to Elio Usai, who invited me for a research stay at the University of Cagliari, Italy in Spring 2023; most of the manuscript was written during these stays. Finally, I thank my beloved wife, Natalia, for her endless patience throughout the time I was nailed to my desk and computer to finish and polish this manuscript, including several weekends and holidays.

May 2023, Grimstad

Author

Michael Ruderman first studied applied physics and then computer and electrical engineering, earning his Dr.-Ing. degree from the Technical University Dortmund, Germany in 2012. He worked in academia for seven years in Germany and two and a half years in Japan at different institutions, and became a faculty member at the University of Agder, Norway in 2015. He has been a full professor there since 2020 where he teaches control theory and engineering in various study programs. He serves on different editorial boards and technical committees and has co-organized several scientific conferences and workshops. His research interests are, among others, motion control, robotics and mechatronics, nonlinear systems with memory, and hybrid and robust control.

1 Introduction

Kinetic friction is just one of the branches of a remarkably large, interdisciplinary, and far-reaching tree that we might call the study of Friction. Without claiming any dominance, this tree has been shaped in many ways by advances in both natural science and engineering, while its roots could be unanimously understood to be in *contact mechanics* and *physics of friction*. While the latter may not itself find a widespread acceptance as terminology, it is directly linked to Tribology, the term suggested first in sixties of the last century for the scientific and engineering subjects related to contact, friction, and wear. The term tribology derives from the Greek verb 'tribo' meaning 'to rub', so with the suffix '-logy' standing for 'study of' one has a well-defined 'study of rubbing' when speaking of Tribology in a broad sense.

Despite multiple notable historical traces of the already ancient scientific ideas and theories about friction as an observable and usable phenomenon, our current understanding of the first verifiable studies of friction is linked to Leonardo da Vinci (1452–1519). He was apparently the first who investigated and formulated the laws of friction quantitatively and left behind a remarkable contribution, including multiple detailed experimental drawings, that allow concluding the two fundamental principles of the classical physics of friction:

- frictional force is proportional to the normal load;
- friction force is independent of the contact area.

Da Vinci was performing multiple experiments, which are similar to those traditionally used in the school's physics like blocks pulled by heavy objects through a rope and pulley system and blocks on inclined plans, and he was also studying various methods for reducing friction like ball bearings and lubrications. As a matter of fact, he was the first to introduce the term coefficient of friction, as a ratio between the resulting friction force and applied normal force, and experimentally determine its typical value of $\approx 1/4$. In his numerous notes, some of which are known as Codex Atlanticus, Codex Arundel, and Codex Madrid, he wrote, for example:

'Each body resists in its slipping with a force equal to a quarter of its weight, the motion being flat and the surfaces dense and clean. ... Bodies that have cleaner surfaces slip easier. In a body with different sides, the friction, made from any of these sides, will not change in resisting to the motion.' [8]; and further he wrote:

'Therefore always when you wish to know the quantity of the force that is required in order to drag the same weight over beds of different slope, you have to make the experiment and ascertain what amount of force is required to move the weight along a level road, that is to ascertain the nature of its friction.' [4].

The results of Da Vinci fell into oblivion for more than 200 years in the middle Ages, and it was only through the seminal works of Guillaume Amontons (1663–1705), followed by Leonhard Euler (1707–1783), and finally by Charles-Augustin

DOI: 10.1201/9781003415015-1

de Coulomb (1736–1806), the study of friction experienced a renaissance, leading to a series of confirmations and refinements of what we understand today under the *law of dry friction*. Euler worked with a mathematical viewpoint on friction and also with experiments, having a basic idea that friction originates from the interlocking between small triangular irregularities and that the coefficient of friction is equal to the gradient of these irregularities. This pioneering interpretation of the rough contact surfaces survived in time and is used also today in the 'Tomlinson Model' of friction on an atomic scale. He considered friction as an effect of the irregularity of surfaces and wrote:

'There is no doubt that friction becomes smaller, the more the surfaces of the slipping bodies are polished and plain, so that there are no more small inequalities, which can hinder the movement' [8].

In Euler's modeling, the bodies in contact were represented as a pair of hilly surfaces with interlocking sawtooth patterns. Although his work on friction does not appear to yield any principally new and relevant results, compared to the findings from the works of Da Vinci and Amontons, he made the first mathematical treatment of an explanatory model of friction, including a theoretical distinction between static and dynamic friction and introducing the well-known symbol for the coefficient of friction, μ.

An outstanding and highly relevant contribution, also for the nowadays understanding of *classical tribology*, was made by Coulomb, whose name is deservingly used for the dry friction law. As a contemporary engineer, he studied separately the static and dynamic friction. Coulomb investigated the dependency of the static friction on four factors, the nature of the materials in contact, the area of contact, the load supported by the surfaces, and the duration of the contact. For dynamic friction, he considered dependency on the first three factors mentioned above and, additionally, on the speed of sliding. He undertook an exact quantitative experimental examination of the dry friction and published his voluminous results in a well-celebrated work 'Theory of Simple Machines', first appeared in 1785 and then republished in the nowadays available new edition in 1821, see [11]. Coulomb used a similar explanation of the origin of friction as Euler, but added also another contribution about friction that is understood today as *adhesion*. Out from his work, the most important concepts have emerged such as surface roughness, asperity deformation, adhesion, and breaking of asperities, cf. also Figure 1.1. As his results and conclusions were well ahead of his time, he already found out that, for instance, the static friction force grows with the amount of time the object has remained stationary. Today, this is often called the *dwell time* in multiple engineering studies. Regarding the physical explanation of friction, Coulomb indicated two possible causes:

'The interlocking of the asperities on surfaces, which can separate only by bending, breaking, or one rising above other; or it should be supposed that between the molecules of surfaces in contact there exists, by their proximity, a cohesion (or adhesion) which should be overcome to produce the movement: only experiments will be able to tell us the reality of these various causes' [8].

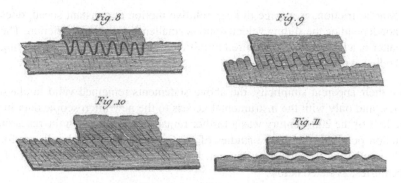

Figure 1.1 Coulomb's representation of interlocking deformable asperities of two rough surfaces in contact [based on Coulomb [11]].

He concluded that, however, only the first cause could explain well the obtained experimental results, and cohesion (adhesion) would only have a small influence. This already recognized, but supposed as only small, influence, that is well explainable by the state of machinery and measurement accuracies of his time, is still of a pivot relevance for the basics of tribology as he wrote:

'We find that friction is about proportional to load, and independent of area; on the contrary, cohesion would act according to the number of the points of contact or to the area of surfaces. We find however that this cohesion is not precisely zero, and we had to take care to determine it in various kinds of experiments' [8].

Therefore, Coulomb was not only confirming and formalizing the Amontons' law

$$F = \mu F_N$$

for the normal load force F_N, but also suggesting a relation of the type

$$F = \mu F_N + \rho A,$$

that is proportional also to the contact area A. He also stated in his work that

'there is another type of resistance, in the friction between surfaces, independent of the load and proportional to the surface areas' [8].

Remarkable is the fact that such behavior is also observed in nano-tribology experiments nowadays, where static friction does not tend toward zero when a load is annulled. Many measurements from the modern friction studies fit with the above relationship proportional to both, the load force and contact area. The above stated contributions have brought friction into the field of applied physics and, over time, crystallized into what is referred to as the Coulomb-Amontons laws of friction. These can be briefly summarized in the following statements:

- Frictional force is independent of the apparent area of contact.
- Frictional force is proportional to the normal load.

- Kinetic friction, as a force to keep relative motion at constant speed, does not depend on the sliding velocity and is smaller than the static friction. The latter is a force which is required to initiate motion between two contacting bodies at rest.

Despite their apparent simplicity, the above statements remained valid in classical tribology, and only with the instrumental access to the new microscopic data in the second half of the 20th century was a further remarkable progress in the research of dry friction possible. Within the studies of friction, however, we should not forget two other pillars of development that emerged in the late 19th and first half of the 20th centuries, as stated below.

First, the developments of Industrial Revolution, including new machinery and also railroads, demanded for study of lubrication. Mineral oils became more used, and the related studies of hydrodynamics and viscous fluids became more and more associated with problems of friction, at least in the mechanical engineering world. The elaborated works of Osborne Reynolds and Richard Stribeck became widely acknowledged and are still in use today. The developed by Stribeck characteristic graph (also known as Stribeck's curve or Stribeck's static model of kinetic friction) distinguish three zones of a sliding velocity, characterizing the resulted thickness of the fluid film and different behaviors of the lubrication and, as consequence, of the kinetic friction force. These zones are the so-called (i) boundary lubrication, (ii) partial fluid lubrication, and (iii) full fluid lubrication.

Second, the so called (in tribology) *minimalistic* models of friction, like the theoretical Prandtl-Tomlinson and Frenkel-Kontorova models, were proposed. The minimalistic models of friction reduces the system to its bare essentials. They can represent a frictional pair by a single particle between two rigid surfaces with asperities. Such models can also describe a system of many interacting or non-interacting particles between two surfaces. Most of the experimental observations are qualitatively recovered by the minimalistic models, cf. [36]. A simple low-dimensional minimalistic model appears useful for understanding qualitatively many physical aspects of the friction. At the same time, minimalistic models provide an intermediate level for describing friction between atomic physics and macroscopic phenomenological approaches, focusing on a small number of relevant degrees of freedom that capture slip motion and describe instabilities during the stick-slip.

One of the main difficulties in understanding frictional response is the complexity of highly non-equilibrium processes, which take place on the contact surfaces and include detachment and reattachment of multiple microscopic junctions between the asperities of surfaces in contact. The friction is, therefore, directly related to instabilities that occur on a local microscopic scale. The well known and investigated, also experimentally, stick-slip phenomena are due to elastic instabilities which play a dominant role in explaining frictional dissipation. One should keep in mind, however, that an interplay between small-scale disorder due to random contacts and elastic interactions between the contacts is a complex problem of statistical mechanics. Still, the theoretical advances in the last decades (reinforced also experimentally with the help of, for example, AMF – atomic force microscopy) allowed significant progress

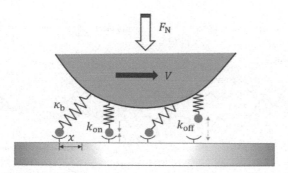

Figure 1.2 Schematic representation of a multicontact model [reprinted from [21]]. The upper body under the normal load F_N is moving with the relative velocity $V(t)$. The randomly distributed contacts are characterized by the average spring constant k_b. The elongation of a contact is x, and k_{on} and k_{off} are the rates of formation and rupture of contacts, which depend on the contact elongation, temperature, and pressure.

in the development of the so-called *multicontact* models; see an example illustrated in Figure 1.2. Two bodies in contact form a set of n random contacts per unit area. These contacts may represent molecular bonds, capillary bridges, asperities between rough surfaces, and for lubricated friction they can mimic patches of solidified lubricant or its domains. Each contact is modeled as an elastic spring connecting the slider and the underlying surface. As long as a contact is intact (unbroken), it is increasingly stretched with a speed equal to the velocity of the slider and thus produces an increasing force that inhibits the motion. After the instability point is reached, a ruptured contact relaxes rapidly to its unstretched equilibrium state [38]. The kinetics of contact formation and rupturing processes (or more generally tribological interactions) depends on the physical nature of contacts and is, therefore, inherently complex and even of a multiphysical nature, cf. Figure 1.3. However, the developed multicontact tribological models enabled a number of key conclusions to be drawn on the mechanism of transitions from the static to kinetic friction in macroscopic systems, cf. [38]:

- An onset of sliding is preceded by a discrete sequence of crack-like precursors – collective modes of the entire ensemble of asperities.
- The transition is governed by an interplay between three types of fronts: sub-Rayleigh, intersonic, and slow fronts.
- A sequence of precursor events gives rise to a highly inhomogeneous spatial distribution of contacts before the overall sliding occurs.

Therefore, the collective behavior of an asperity ensemble, that composes a frictional interface, determines the transition mechanism from static to kinetic friction.

Despite all the meritorious achievements of modern tribology, including nanotribology of the last decades, solid friction and kinetic friction represent a case that also requires an analysis at a *mesoscopic* level to explain and understand frictional

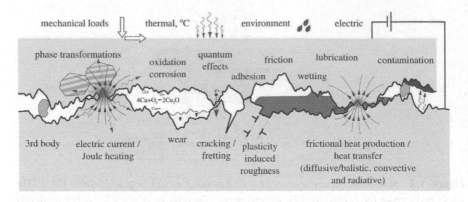

Figure 1.3 A schematic representation of multiphysical nature of tribological interactions: two different solids with rough surfaces and relevant material microstructures are brought into mechanical contact and exposed to various loads: mechanical, thermal, electric, and environmental [reprinted from [37]].

phenomena. In fact, research has shown that friction at the atomic scale presents some marked differences compared to the properties observed in macroscopic friction. Adhesion plays a very significant role. In many cases, static friction does not tend towards zero when the load is annulled but persists for negative charges, up to a limit after which there is a sudden detachment of the surfaces. This behavior appears in contradiction with what is observed at macroscopic scale. However, it can be explained by the presence of irregular asperities of surfaces. Since the asperities have various dimensions and deformations, when the load is reduced, the junctions break or detach one after the other, so that, when the load is annulled, the real contact area becomes almost zero, and therefore the friction force tends to zero and the phenomenon of sudden detachment disappears [22]. Moreover, it was recognized that the atomic scale would be proper for the description of some friction processes, but not others where the dissipation of mechanical energy is regulated by mesoscopic granularities of dimensions from 0.01 to 10 micrometers, which is typical of polycrystalline materials. The atoms within a single grain occupy an ordered crystalline lattice, but the material is disordered at the macroscopic scale [8]. The formation and the propagation of crystalline defects among the grains would constitute the primary mechanism for inelastic deformation and also hysteresis in friction. In the common sense of modern tribology, it seems that the problem of an atomic explanation for friction is not yet resolved, even if enormous progress has been made over the last decades. For instance, according to the former fundamental investigations of tribology with AMF:

'Despite the heavy volume of work performed so far, there does not exist a fundamental understanding of tribological processes (although a great deal of practical successes have occurred). It would be fair to say that friction itself is one of the most common, yet least understood, physical phenomena.' [9].

Coming back to a mesoscale of friction, which is certainly pivotal for many engineering problems, an intermediate level can be assumed as a necessary means of relating 'micro' and 'macro' tribology. An advantage of this intermediate level is that, unlike molecules and atoms, the mesoscopic elements keep the essential properties of macroscopic objects, and can therefore be treated as small objects that have temperature, density, elasticity, can be deformed and can interact through contacts, cf. [8].

Beyond the mesoscopic multicontact models mentioned above, cf. Figure 1.2, the so-called phenomenological *rate-state* (RS) approaches of interpreting and modeling the friction on mesocale were proposed starting from early eighties. At the largest macroscopic scale, phenomenological RS models simplify the description of friction, introducing only one or two dynamical equations with coefficients chosen to fit experimental quantities and then used to describe a wide range of observed frictional behavior. Here belong the transition between stick and slip (regular or chaotic), smooth sliding, and variations of friction for a sudden change of velocity, cf. [38]. Often, the RS models are even the best available approaches to describe macroscopic friction in the ordinary world, from the micro-size to larger scales. However, most of the 'state variables' in RS models cannot be easily related to the physical system properties, a fact that often limits the insight and reasoning of these models. The appearance of RS models is commonly associated with the works by Andy Ruina [35], [25]. A general empirical rate-state-dependent friction law was proposed in the Ruina's model, based on experiments involving a suddenly imposed step change in the velocity. The experiments by Ruina suggested that the friction force response to the velocity step is a combination of an instantaneous increase with the velocity and a first-order-like decay with the evolving state. This effect has been widely recognized by other RS friction models and is often referred to as *friction lag*, which describes the rate-dependent transient behavior of the gross sliding friction. The phenomenological RS models of friction were also suggesting that an interaction between the two surfaces occurs via both elastic and plastic contacts. From the standpoint of tribological processes, the junctions of a contact pair have two important behaviors: they deform elastically, giving rise to pre-sliding displacement; and both the boundary contact layer and asperities deform also plastically, giving rise to static friction. According to RS approaches, as proposed by Ruina, the dependence of the friction force on slip history is described by an (experimentally motivated) constitutive law where the friction force is dependent on slip rate and state variables. The state variables are defined macroscopically by evolution equations for their rates of change in terms of their present values and slip rate [35].

The varying stiffness of contact 'springs', mimicking the asperity contacts, became a typical characteristic of RS modeling approaches, that makes also the concept of rate- and state-dependent friction indistinguishable between static and kinetic friction. Moreover, the interaction of asperities and their kinetics have sometimes also been linked to those of the bristles of a brush. The physical paradigm underlying a *bristle model* is a pair of facing surfaces with deformable bristles extending from each surface, where the friction interaction occurs at the bristle tips. This simplified, though well understandable and quite flexible, interpretation allows for the inclusion of such tribological interactions as elastic and plastic deformation, adhesion, ploughing of surfaces into each other (under the normal load), formation of junctions, and others. Another relevant feature of RS approaches is a characteristic pre-sliding range in which the motion occurs accompanied by a deformation of asperities, correspondingly their rupture and detachment, in the sliding interface. Comparing to the multicontact tribological models, the pre-sliding range can also be brought into association with elongation of the single contacts, cf. Figure 1.2, and thus regions of local elastic instabilities before a slip motion sets on. Ahead of the RS approach proposed by Ruina, it was Philip Dahl who introduced a simple phenomenological dynamic friction model, also based on an internal state of the friction system. The Dahl model imposed hysteresis in the pre-sliding displacement and provided an answer to the problem of jump discontinuity for the Coulomb friction. Further developments of the phenomenological RS approaches to solid and dynamic friction followed, often driven by the purely engineering studies.

By the end of the above excursion, we should keep in mind that a RS modeling is, despite its deceiving simplicity, extremely useful (especially in engineering) for understanding many aspects of the static and kinetic friction. At the same time, its extreme success tends, however, to hide the actual complexity of frictional phenomena.

1.1 FURTHER READING

Although the following course materials are written to be self-contained and should enable readers to understand the key principles and methods of the analysis and design presented, a more interested audience may find it useful to consult also additional literature. This is listed below in a certain association with the content and focus of the following chapters.

A historical overview of the scientific models and theories of studying friction, divided into four phases starting from Leonardo da Vinci and Coulomb and going until nanotribology and friction at the atomic scale of nowadays, can be found in [8]. Further details on discovering the dry friction law, especially in relation to the works done by Coulomb and Euler, can be found in [39]. Also in [23], the contributions of Coulomb and Amontons to the generalized laws of friction are summarized together with some historical aspects. A physics oriented colloquium, with an extensive referencing therein, on the friction modeling from nanoscale to mesoscale can be found in [38]. From an engineering perspective, the tribological effects which are more specifically related to the Stribeck phenomenon can be consulted, for instance, in the Stribeck memorial lecture in [14]. Also the Dahl friction model, as one of the first and most successful attempts to bypass the Coulomb friction discontinuity, can be studied in detail in [10]. For a deeper mathematical understanding of the system dynamics with Coulomb friction discontinuity, which means of the differential equations with discontinuous righthand side and of what is called solutions in *Filippov sense*, the reader is referred to [12]. A compact introductory physical insight into the nonlinear nature of friction, adhesion, lubrication and wear of surfaces during the relative motion can be found in [36]. For those readers who will find time and inspiration to immerse further into frictional studies, then on the level of surface mechanics, material science, and tribology, the author would like also to recommend two textbooks. The first one is *Friction and Wear of Materials* by E. Rabinowicz [24]. The second one is *Sliding Friction: Physical Principles and Applications* by B.N.J. Persson [22].

For studying frictional phenomena in servo mechanisms and drives, an essential and well-celebrated survey can be found in [5]. A later seminal work, which also deals with nonlinear friction forces and complex friction force dynamics, is provided in [2]. Some discussed details about the kinetic friction and contact surfaces of a motion system can be found in [33]. For studying experimental characteristics of the kinetic friction, some practical tribological investigations with use of a tribometer setup designed for the macroscopic measurements can be looked up in [15].

For basics and details on the Lagrangian mechanics, also on the Rayleigh dissipation function, the comprehensive treatment in [18] is recommended. Also the well-illustrated examples for deriving the Newton-Euler and Lagrange dynamics in robotics and mechatronics can be found in [19]. More details on the break-away forces are provided in [26]. For further discussion of how the motion reversals and stop are governed by the pre-sliding friction behavior, the reader is welcome to look in [34]. A discussion on energy balance and dissipation at motion reversals can also be found in [28]. The details on invariant set of trajectories of the feedback controlled motion with Coulomb friction are given in [27]. Also some interesting experimental observations of long-term stick-slip cycles of a feedback controlled motion with Coulomb friction can be found in [7]. The associated mathematical problems of differential inclusions can be found in [16], deepened in the theory.

For general topics of feedback control design, including output and state feedback controllers, the reader is referred to, for instance, [13]. A more detailed discussion of the output relay feedback compensation of the Coulomb friction can be found in [3] and [30]. The associated details of harmonic balance analysis for two-relay feedback systems can be looked up in [1], while reference is made to [6] for the fundamentals of describing function methods. The presented observer-based friction compensation is initially provided in [31], while an advanced and simplified form of the asymptotic observer of nonlinear friction is discussed in detail in [29]. For the basics of asymptotic linear state observers, often referred to as Luenberger observers, an interested reader is welcome to look back at [17]. A former summarizing work on the friction models and friction compensation, also including an observer-based approach, can be found in [20]. Another former book on the control of machines with friction [4] can also be consulted.

1.2 NOTATION

Unless stated otherwise, we will use standard mathematical notations, some of which are worth recalling, as correspondingly stated below.

The first- and second-time derivatives of a variable y are denoted by $\dot{y}$ and $\ddot{y}$, respectively. The likewise used notation for derivative of y with respect to x is dy/dx, while $\partial y / \partial x$ denotes the partial derivative of y with respect to x. The time argument is denoted by t, the complex Laplace variable by s, the angular frequency by ω, and the imaginary number of complex numbers by j. The system constants are denoted by either lowercase letters or capital letters with a subindex, for example, m, k, σ or C_f, K_p, K_1. The functions and functionals are denoted by either lowercase Greek letters, like α, κ, or by capital letters with the corresponding argument, for example, $f(\dot{x}), H(s), V(x,t)$. The constant matrices and vectors, on the contrary, are denoted by single capital letters without argument, for example, A, B, L. The discontinuous sign operator is defined as constant for positive and negative numbers and in the Filippov sense on the closed interval in zero, so that

$$\text{sign}(y) = \begin{cases} 1, & y > 0, \\ [-1,1], & y = 0, \\ -1, & y < 0. \end{cases}$$

2 Phenomena of Kinetic Friction

2.1 KINETIC FRICTION INTERFACE IN MECHANISMS

To understand principles of a friction interface in drives and mechanisms used in the robotic and mechatronic systems, we need first to consider a frictional pair of two mechanical bodies in contact. Typical topology of a rough surface pair in contact can be represented as depicted in Figure 2.1 (left), cf. with [33]. Despite an irregular landscape of the rough surfaces in contact, a particular surface topology can be characterized by some average values of the height, distance, distribution, and elasto-plastic properties of the surface asperities. Under the action of normal load exerted on the upper body and its own weight, the upper surface penetrates into the lower surface, as a result of which the so-called furrows and junctions are formed. Then, when the upper surface is forced to translate tangentially against the lower surface, the interlocking asperities begin to counteract, causing the tangentially acting friction forces to develop on the interface of a contact pair. In the course of this process, the furrows and junctions can undergo either plastic micro-deformations, changing their shape and leaving traces, or transient elastic deformations, with subsequent restoration to the original shape. This is accompanied by losses of energy in the form of structural damping and, at long last, heat dissipation. It is important to emphasize that considering the lumped modeling of kinetic friction, we need to focus only on a *point contact*. This way, neither a distributed continuum of the contact points nor decomposed rolling friction need to be covered by the modeling framework utilized in the following. We will also purposefully neglect another well-known frictional phenomenon, namely the static (i.e. stiction and adhesion) friction force and the so-called dwell time – the resting time of a contact pair during which the adhesion between both surfaces is established.

An interaction between the asperities of a rough surface pair can also be interpreted by the so-called bristle model, in which the distributed bristles of a brush mimic the asperities deformation (elastic or plastic) during a tangential displacement. Despite a rough landscape of both surfaces, cf. Figure 2.1 (left), for the sake of simplicity one can assume the only upper surface to be deformable and moving along the rigid lower surface, as schematically represented in Figure 2.1 (right). For nonzero relative velocities, i.e $\dot{x} \neq 0$, a continuous elastoplastic deformation of the contact layer of the upper surface results in the resistive kinetic friction force. Note that the kinetic friction force here can include both the dry and viscous components, the latter becoming more pronounced when additional lubricant is applied on the interface. For very low relative velocities, a *boundary lubrication* regime predominates, where the lubricant's penetration and distribution over the landscape of a rough surface can reduce the static friction, correspondingly the effective Coulomb friction, and

DOI: 10.1201/9781003415015-2

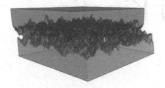 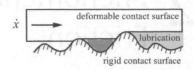

Figure 2.1 Typical topology of the rough surface pair in contact (left), schematic representation of a deformable upper surface moving along the rigid lower surface with lubricant's penetration between the surface asperities (right).

therefore attenuate the stick-slip effects. However, the velocity here is not sufficient to build up a separating fluid film between the contact surfaces. With an increasing relative velocity, *partial fluid lubrication* occurs, leading to a larger separation between both surfaces. The behavior of a partial fluid lubrication can be better understood by certain analogy with a water skier. At zero velocity the skier is plunged into the water and solely supported by buoyancy. At some critical velocity the skier becomes hydrodynamically supported by her or his motion. Apparently, there is a range of velocities between floating and skiing, where the skier is partially hydrodynamically supported. These velocities are analogous to the regime of partial fluid lubrication. As the partial fluid lubrication increases, the number and frequency of asperity junctions decrease, resulting in a temporary reduction of the friction force. This well known by-effect of the viscous friction is mostly denoted (especially in engineering world) as Stribeck behavior, or Stribeck weakening curve of decreasing friction with an increasing relative velocity. We mention, however, that these effects stay outside of our main focus in the following. With further increase of the relative velocity, *full fluid lubrication* sets on, which is mainly determined by hydrodynamics of the viscous fluids that make up the lubricating medium. According to the Reynolds' general theory of hydrodynamic lubrication, the hydrodynamic pressure and the shear stress in the lubricating film increases with an increasing velocity. This results in a viscous resistance, or linear drag (i.e. Stokes' drag), which gives rise to the force of drag approximately proportional to the relative velocity. This effect is well known as a standard viscous friction, cf. later with section 2.3.

Now, more specifically for 1DOF mechatronic systems, let us exemplify a friction interface for electromechanical drives equipped with the ball screw transmission, as schematically shown in Figure 2.2 (left), cf. with [32]. Such actuating mechanisms provide the motor speed reduction, accompanied by transformation of the rotational motion into the linear one, and corresponding amplification of the drive force through the ball screw gearing, which principal structure is shown in Figure 2.2 (right). Worth noting is that the ball screw drives are widely used in different industries, targeting for example precision applications in the CNC (computer numerical control) machines, linear robotic axes, instrumentation and measurement devices, and others. When a screw shaft, driven by a controlled electric motor, is rotating, the steel balls

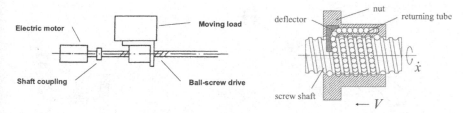

Figure 2.2 Schematic representation of 1DOF drive system with ball screw (left); principal structure of a ball screw transmission (right).

travel by rolling along the grooves of the screw shaft and the ball nut. The latter is integrated into a stiff housing (cf. with moving load part in Figure 2.2 (left)), which supports the payload and, thus, constitutes the output unit of a ball screw transmission. When arriving at the end of the nut, the balls are passing back through the so called returning tube and, then, the rolling circulation along the grooves repeat. Whenever the nut strokes the screw shaft, the balls repeat the same circulation, which is mechanically constrained in a closed kinematic chain, once by the grooves of the screw and nut, and once by the returning tube. This way, the overall contact surface of a ball screw transmission is almost the same, while a constant number of balls rolling under the loaded contact gives rise to the overall counteracting friction. It is worth noting that although the intensity of the friction force varies with the state of contact of the balls, the total rolling contacts still reduce the friction, comparing to friction of the otherwise also existing linear sliding guides. During the manufacturing assembly, the ball screws are usually preloaded, both to minimize a further elastic deformation (of the balls) owing to the external payload forces, and to eliminate axial play and thus backlash between a screw shaft and a ball nut. With the above in mind, and neglecting an additional but inferior friction in the motor shaft bearings, a ball screw drive can be associated with a lumped-parameter kinetic friction owing to the overall contact interface between the balls and grooves. Despite this generally favorable simplification, one should keep in mind that large variations in the normal payload or external factors such as ambient temperature or lubrication and dust in the ball screw can affect the resistive friction forces. This can necessitate an additional identification of the friction model parameters in the system.

In contrast to 1DOF drive systems, as exemplified above, the robotic manipulators (or mechanical manipulators in general) can have multiple joints and corresponding drives, see Figure 2.3 (a). Regardless whether prismatic (P) or rotary (R) joints are in place to actuate the manipulator's own body and additional payload, the friction behavior in the drives can change depending on the instantaneous configuration of the joints. For an example illustrated in Figure 2.3 (a), the position state in the rotary joint coordinates (x_2, x_3) will dynamically change the overall distribution of the moving masses and, thus, also the normal load in the respective drive units. However, these changes are mostly bounded by the kinematic and mass constraints of a particular manipulator. The kinematic constraints will determine in which way and how far

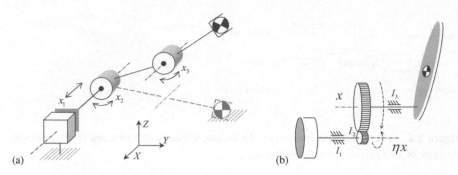

Figure 2.3 Kinematic structure of a PRR-manipulator with one prismatic (P) and two rotary (R) joints (a); schematic representation of a rigid robot joint with gearing (b).

the certain axes will change their orientation and, therefore, configuration-dependent normal load, cf. the P-axis in x_1 coordinates and R-axes in x_2 and x_3 coordinates in Figure 2.3 (a). The mass constraints, on the contrary expressed in terms of the own weights, center of gravity of the linking bodies, and the maximum permissible payload of manipulator determine an upper bound of the normal load for each robotic joint. Considering a principle structure of the rigid robot joint with gearing, as schematically shown in Figure 2.3 (b), it is easy to recognize that at least three frictional interfaces, denoted by I_1, I_2, I_3, will contribute together to the overall friction torque of the joint. The first one I_1 is due to the bearings of an actuator (mostly electric motor), which provides the angular displacement in the relative ηx coordinates, where η is the nominal gear ratio. The second one I_2 is directly associated with teeth meshing of the gear unit embedded into the robot joint. Here it should be noted that the complexity of frictional contacts can vary depending on the type of gear in use, for example, strain wave gearing or epicyclic (i.e. planetary) gearing or cycloid gearing. The third frictional interface I_3 can (but not necessarily should) be associated with bearings of the output axis of the robot joint unit. Despite this apparently complex and superimposed structure of mechanical contact interfaces, the overall friction force (correspondingly torque) of a robot joint with one generalized DOF denoted by x_i can be assumed as a superposition

$$f_i = \sum_{k=1,\ldots,3} f_{I_k}, \tag{2.1}$$

of friction forces occurring at the single interfaces. Indeed, for an induced actuator torque to start moving the output joint axis, i.e. for $\dot{x}_i \neq 0$, it must first overcome the friction torques at all three contact interfaces as in Figure 2.3 (b).

2.2 APPEARANCE AND SIGNATURE OF FRICTION EFFECTS

An inherent problem of studying kinetic friction in the mechanisms and drives is the fact that friction forces are not directly measurable. Quite opposite to inertial forces and contact forces (or more generally interaction forces), the friction forces are lacking a suitable interface at which a sensing element could be appropriately applied. We recall that the inertial forces can be captured by a sufficiently accurate measurement of acceleration, provided the inertial mass is known precisely, that is often the case. Also a variety of force and torque sensors can be applied either on a contact interface with environment or between the mechanical parts of a mechanism, for which an interaction force should be determined. A classical example represents the torque sensor integrated into the drive chain with a common rotary axis. This makes possible to determine the torsional torque and, thus, the rotational stiffness of interconnected elements in a drive. In contrast, the friction forces occurring at the contact interface, which is a structurally engaged but not a fixed one, are tangential to the surfaces and orthogonal to the normal loads. This spatial configuration makes them almost impossible for a direct measurement and would require (even only in theory) a sensor element itself to become a contact surface between both rubbing bodies. Therefore, the measurement of friction forces can mostly be claimed only based on an accurate measurement of other forces in a static or dynamic equilibrium. Out from that equilibrium, the value of friction force f^{fric} can then be calculated by an appropriate decomposition. This can be written, in general terms, as

$$f^{\text{fric}} + \sum_i f_i^{\text{meas}} = 0, \qquad (2.2)$$

where f_i^{meas} are the measurable forces in a force-balanced system. Examples of the associated devices and apparatus, purposefully designed for accurately measuring the kinetic friction, are rare and associated mostly with some specific laboratory equipment differing from an everyday engineering practice.

Against the above background, a detection and subsequent identification of friction effects in the drives become a non-trivial task for system and control engineering. While the viscous friction components are well inline with a linear systems paradigm, here associated with linear system damping and thus with identification in both time and frequency domain, the nonlinear frictional effects pose a serious challenge for being properly isolated within the overall system response. In the following, we will demonstrate this issue based on some, so to say, hidden and equivocal signatures of friction. Assume a most simple first-order drive system described by

$$m\ddot{x}(t) + f(\dot{x}(t)) = u(t), \qquad (2.3)$$

where an inertial mass m is actuated by the input force u, and the counteracting kinetic friction force is $f(\dot{x})$. Also assume a frequent situation where the available input and output signals are the drive force $u(t)$ and the relative velocity $\dot{x}(t)$. Now let us distinguish three different types of kinetic friction $f(\cdot)$, which are then used for analyzing numerical response of the system (2.3) and for accentuating the problems of a hidden signature of frictional effects.

(A) Kinetic friction in (2.3) is a purely linear viscous, cf. further with (2.5).
(B) Kinetic friction in (2.3) is a combination (2.7) of linear viscous and Coulomb friction. The Coulomb friction is discontinuous as in (2.4).

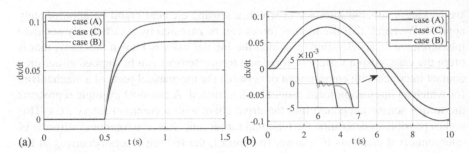

Figure 2.4 Time-domain response of the drive system (2.3) with three cases (A), (B), and (C) of kinetic friction: the step response in (a) and harmonic response in (b).

(C) Kinetic friction in (2.3) is a combination (2.7) of linear viscous and Coulomb friction. The Coulomb friction is continuous and described by the Dahl friction model.

Note that in the last case (C), the Dahl friction model with only two parameters, one of which is the Coulomb friction coefficient, cf. with (2.4), is assumed for the sake of simplicity. That means also other (dynamic) friction models which at least avoid discontinuity of the Coulomb friction at zero velocity crossing could equally be assumed for the case (C), cf. later with section 3.2.

In order to examine what can be said about the presence of kinetic friction in the drive system (2.3), consider first the time domain response of the induced relative velocity $\dot{x}(t)$. Such response, once to the step and once to the sinusoidal input, are exemplary shown in Figure 2.4 (a) and (b) correspondingly. It is visible that the single step-response does not allow to distinguish between the Coulomb friction cases (B) and (C). Also the case (A) does not allow for concluding unambiguously whether a linear or nonlinear friction is acting in the drive system (2.3). Indeed, the shape of the transient response are very similar to each other, and only the steady-state value distinguishes the case (A) from (B) and (C). Here it is important to recall that during identification, the steady-state level is given by the recorded measurement only, so that a distinction between (A), (B), and (C) based on the single step response measurement becomes impossible. When inspecting the harmonic response of the drive system (2.3), see Figure 2.4 (b), the difference between the purely viscous friction and the kinetic friction which includes the Coulomb terms becomes more pronounced. Cases (B) and (C) differ from (A) by a visible dead-zone phase, which becomes larger, a lower input frequency is applied. Still, a certain but minor difference between (B) and (C) is only visible during the motion stop, i.e. at the beginning of dead-zone phase, see zoom-in in Figure 2.4 (b). It is also worth emphasizing that in a real physical drive system, the measured $\dot{x}(t)$ signal will mostly be affected by the unknown, and thus not decomposable, process and measurement noise. This fact will make a distinction between the cases (A), (B), and (C) even more unrealistic, thus making such time domain identification highly uncertain for the systems with friction.

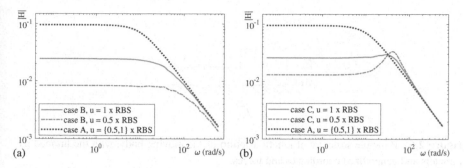

Figure 2.5 Frequency response of the drive system (2.3) with three cases (A), (B), and (C) of kinetic friction: comparison between the input excitation by 1×RBS and 0.5×RBS.

For obtaining a more pronounced signature of kinetic friction within the dynamic system response, one can apply a broadband excitation to the drive system (2.3) and identify the frequency response function (FRF), i.e. $H(j\omega) = \dot{x}(j\omega)/u(j\omega)$. Recall that while an FRF (equivalent to Bode diagram) can be calculated analytically for the case (A), which is linear and has a closed-form of the transfer function, it is not possible for the cases (B) and (C) in this way. However, an FRF can be pointwise determined from the recorded input-output data and, afterwards, smoothed over the frequency range of system excitation. Applying a broadband excitation in form of a random binary signal (RBS), the $H(j\omega)$ values are obtained also for the cases (B) and (C). Comparing the FRFs, exemplary shown in Figure 2.5 for two levels of the input signal 1×RBS and 0.5×RBS, one can recognize that the amplitude response $|H(j\omega)|$ changes significantly in both cases of nonlinear friction. While the linear case (A) remains unchanged, due to the superposition principle and therefore invariance to the input amplitude, the gain characteristics of the case (B) decreases with a decreasing level of the input excitation, cf. Figure 2.5 (a). In case (C), which incorporates a simple yet dynamic friction model, the transfer characteristics experience even a resonant behavior, cf. Figure 2.5 (b), which is associated with the initial stiffness of frictional contacts. Recall that the frictional contacts behave similar to an oscillating structure at certain higher frequencies, when the input excitation becomes insufficient for generating a macroscopic motion and, thus, for breaking away from the pre-sliding regime, cf. later with section 3.2. At large, from Figure 2.5 one can recognize that the kinetic friction is not only shaping the dynamic response of the drive system (2.3) but also making it both frequency- and amplitude-dependent. This fact can have a significant impact on the system identification and, therefore, also on the resulting performance of feedback control systems to be designed.

2.3 ASSUMPTIONS FOR KINETIC FRICTION IN DRIVES

Before we can analyze the dynamic behavior of systems with friction, which will be our main focus in the subsequent chapter, several basic assumptions should be made about the frictional phenomena we are dealing with. This will serve not only for

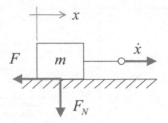

Figure 2.6 Principle structure of kinetic friction of the moving body, with the lumped parameters and generalized coordinates and forces.

better demarcating which friction effects we consider and which we do not, but also for specifying the modeling framework that we assume as sufficient for describing the friction forces during our analysis and control design. The principal mechanical structure, in the generalized $(x, \dot{x})$ coordinates, can be assumed as shown in Figure 2.6. When it comes to the generalized coordinates and forces, all types of common 1DOF drives can be viewed in such unified way, while the linear displacements and forces are understood for translational motion, and the angular displacements and torques for rotational motion. The lumped inertial mass m gives rise to a constant load, expressed by the normal force F_N, while the inertial body itself rests on a flat and homogeneous but, at the same time, rough surface. Either a forced or free relative motion is considered with only one degree of freedom, so that the drive body experiences the relative displacement with $\dot{x} \neq 0$, as long as not in the idle state. As a result of rubbing surfaces under the normal load, the tangential friction force F arises, which is directed oppositely to the relative velocity $\dot{x}$. Also important to recall is that a simplified point-contact is assumed and, therefore, there is no spatial distribution of contact forces over contact area of the moving body. With this in mind, two first relationships valid for the kinetic friction can be qualitatively expressed as (i) $\text{sign}(F) = -\text{sign}(\dot{x})$, and (ii) $(\dot{x}, F_N) \mapsto F$. Note that (i) is written in the opposite coordinates of relative motion and reactive force, and (ii) constitutes a functional map. Then, the following basic assumptions can be made as sufficient for our subsequent discussions.

1. The Coulomb friction force at steady-state motion depends on the motion direction only and is given by

$$F_c = C_f \, \text{sign}(\dot{x}). \tag{2.4}$$

The nonlinear friction law (2.4) is parameterized by the Coulomb friction coefficient $C_f > 0$ and represents the most simple modeling approach with discontinuity at the velocity zero crossing.

2. The viscous friction force depends linearly on the relative displacement rate only and is given by

$$F_v = \sigma \dot{x}. \tag{2.5}$$

The linear friction law (2.5) represents the standard viscous damping and is parameterized by the viscous friction coefficient $\sigma > 0$.

3. Both the Coulomb and viscous friction coefficients are generally weakly known and uncertain, correspondingly they can be time-varying. The rather fast variations can appear as position- or load-dependent, while slower variations can be due to multiple environmental factors like, for example, temperature, state of lubrication, wear, surface cleanness, dwell time and adhesion. Taking into account the above reasoning, the following can be written

$$C_f, \sigma \neq \text{const}, \quad 0 < C_f < C_f^{\max}, \quad 0 < \sigma < \sigma^{\max}, \quad (2.6)$$

where $C_f^{\max}, \sigma^{\max} > 0$ are the known upper bounds.

4. The total kinetic friction at steady-state is a superposition of the Coulomb and viscous friction forces, given by

$$F(t) = F_c(t) + F_v(t). \quad (2.7)$$

The superposition (2.7) is consistent with the established approaches of modeling systems with friction, especially when using the Newton-Euler dynamic equations, and also when incorporating additionally the Stribeck effects. The superposition principle can be temporary lost during the dynamic transients, particularly at the reversals of motion, where the viscous friction effects can subside and also lose their linearity.

5. The frictional transients, i.e. where $\dot{F}(t)$ becomes nonconstant and too essential for being neglected, are characteristic of either the onset and stop or reversal of the relative motion. The friction lag of $\dot{F}(t)$ during gross sliding will be, however, purposefully neglected.

3 Motion Dynamics with Friction

3.1 MOTION EQUATIONS AND ENERGY DISSIPATION

For analyzing the system dynamics with friction, one needs first to look into the fundamental equations of motion, like those formulated in the Lagrangian mechanics. More specifically here, we want to look at the Lagrange's equations of the second kind. Since the number of such equations is equal to the number of generalized co-ordinates, i.e. the number of degrees of freedom in holonomic systems, we restrict ourselves to the scalar case in all the following discussions. Indeed, as one degree of freedom of a drive system with friction is assumed, cf. with chapter 2, all generalized coordinates and forces become automatically one-dimensional. We recall that the Lagrangian

$$L(x, \dot{x}, t) = T(x, \dot{x}, t) - V(x, t) \tag{3.1}$$

is the difference between the total kinetic $T(\cdot)$ and potential $V(\cdot)$ energies of the system. The Lagrange's equation (of the second type) is then given by

$$\frac{d}{dt}\frac{\partial L}{\partial \dot{x}} - \frac{\partial L}{\partial x} = U, \tag{3.2}$$

where U summarizes the external, important to emphasize *non-conservative*, forces in a Lagrangian system. When moving the $\partial L / \partial x$ term, which has the dimension of a force, to the righthand side of (3.2), and recalling that $\partial L / \partial \dot{x}$ represents a momentum, one can write

$$\frac{d}{dt}\frac{\partial L}{\partial \dot{x}} = \frac{\partial L}{\partial x} + U \quad \Rightarrow \quad \frac{d}{dt}\text{momentum} = \sum \text{forces}. \tag{3.3}$$

It can easily be seen that what we declared on the righthand side of (3.3) is just a restatement of the Newton's law in the generalized coordinates. Furthermore, it is visible that for an autonomous conservative system, i.e. with $U = 0$, the kinetic and potential energies are well balanced through

$$\frac{d}{dt}\frac{\partial L}{\partial \dot{x}} = \frac{\partial L}{\partial x},$$

and the dynamic system behaves as lossless.

Let us analyze the above mentioned fact of lossless by using an illustrative example of the second-order drive system, which consists of an inertial body (parameterized by m) connected to the ground through a return rotary spring (parameterized by stiffness k), see Figure 3.1 (a). The kinetic and potential energies of the system are

$$T = \frac{1}{2}m\dot{x}^2, \quad V = -\int_0^x (-kx)dx = \frac{1}{2}kx^2, \tag{3.4}$$

DOI: 10.1201/9781003415015-3

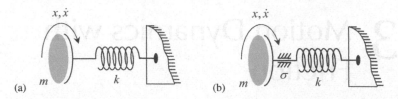

Figure 3.1 Second-order lossless (a) and frictionally damped (b) drive system of an inertial disk on a rotary spring.

respectively. Taking the associated Lagrangian (3.1) and evaluating the Lagrange's equation (3.2) one obtains the dynamics equation of that autonomous system as

$$m\ddot{x} + kx = 0. \tag{3.5}$$

It is not difficult to see that the dynamic behavior of our above example is nothing but an (undamped) harmonic oscillator, whose periodic trajectories depend on the initial values $x(0), \dot{x}(0)$ only. Now, let us convince ourselves why the autonomous system (3.5) is lossless, and that from an energy conservation perspective. Differing to the Lagrangian (3.1), we will now write the total system energy as $E(t) = T(t) + V(t)$ and, then, evaluate its time derivative as

$$\dot{E} = m\dot{x}\ddot{x} + kx\dot{x} = \dot{x}(m\ddot{x} + kx) = 0, \quad \forall \quad (x, \dot{x}) \in \mathbb{R}^2. \tag{3.6}$$

The above equality to zero is evident since the expression in the brackets, on the righthand side of (3.6), is the expression of system dynamics itself, which is equal to zero according to (3.5). Since the time derivative of the total energy is zero, the system behaves conservatively, and there is neither energy inflow nor energy losses for all times t and in the entire state-space $(x, \dot{x}) \in \mathbb{R}^2$. This apparently natural realization for a harmonic oscillator is still interesting to discuss further with regard to the relationship between the state trajectories and the energy conserved in the system (3.5). The total energy E of the system (3.5) can be rewritten as

$$\frac{x^2}{2Ek^{-1}} + \frac{\dot{x}^2}{2Em^{-1}} = 1, \tag{3.7}$$

which obviously represents an equation of the ellipse. Both axes of the ellipse (3.7) depend on the system parameters, but also on the constant energy E conserved in the system as visible from Figure 3.2. The amount of energy, stored in the lossless autonomous system through non-zero initial conditions $x(0), \dot{x}(0)$, is directly reflected in the size of ellipse and, therefore, amplitude of the periodic $x(t), \dot{x}(t)$ solutions.

Now, we can greatly fall back on the fact that nearly every physical drive system, unless it is permanently supplied by control, will continuously lose energy and, in other words, dissipate it via either frictional or structural damping. Indeed, an autonomous system (3.5) is equivalent to a *perpetuum mobile* if neither damping nor control are included. Since the frictional damping effects are of our prime interest, we can extend the previous example of a drive system by the mechanical bearing

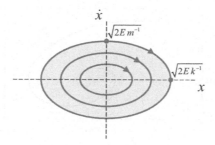

Figure 3.2 Phase portrait of the state trajectories of the system (3.5) for different E values.

as illustrated in Figure 3.1 (b). When assuming here a viscous damping only, which is parameterized by the linear coefficient σ, the system dynamics (3.5) has to be extended correspondingly. One can recognize that this extended drive system has a classical mass-spring-damper configuration, so that its dynamics can directly be obtained by, for example, Newton-Euler approach of a free body. Cutting free the body depicted in Figure 3.1 (b), and respecting the sign of the inertial force $m\ddot{x}$ and the counteracting spring and damper forces kx and $\sigma\dot{x}$, respectively, one can write

$$\sum \text{forces} = \text{inertial force} + \text{damper force} + \text{spring force} = 0,$$

according to the Newton's second law. Writing down, with regard to the above, the corresponding forces which are in superposition we obtain

$$m\ddot{x} + \sigma\dot{x} + kx = 0. \tag{3.8}$$

The above expression describes the well-known linearly damped second-order oscillator. It is also well known that if all roots of the corresponding characteristic equation

$$ms^2 + \sigma s + k = 0$$

have the negative real part, that means they are lying in the complex left-half plane, the system (3.8) is said to be globally asymptotically (and also exponentially) stable. Still, we would like to look also on what happens with the system energy, similar as it was done before for the system (3.5). Taking the time derivative of the total energy E of (3.8), one obtains

$$\dot{E} = \dot{x}(m\ddot{x} + kx) = -\sigma\dot{x}^2 < 0, \quad \forall \ \{(x,\dot{x}) \in \mathbb{R}^2 \,|\, \dot{x} \neq 0\}. \tag{3.9}$$

The above inequality means that only when a state trajectory crosses x-axis in the phase plane, the $\dot{E}$ value becomes temporary zero. Here we purposefully stress that this is temporary, since the vector field at $\dot{x} = 0$ is $\ddot{x} = -m^{-1}kx \neq 0$ everywhere except the origin. Following to that, a state trajectory will never stay on the x-axis and, thus, will converge unavoidably to the origin. Hence, the origin is an unique and stable equilibrium of the system (3.8). An example of such converging trajectories of

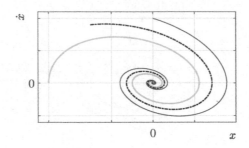

Figure 3.3   Phase portrait of the state trajectories of system (3.8) for different initial values.

(3.8), with a conjugate-complex pole pair and thus oscillating behavior, is illustrated in Figure 3.3 for the different initial conditions $x(0), \dot{x}(0)$.

Once we have seen how the linear frictional damping is entering the motion dynamics (3.8), and that by applying the Newton-Euler approach, it is natural to seek also for a general Lagrange's analogy, in order to include the dissipative terms. This would allow incorporating both, the viscous and Coulomb friction into the energetic balance, correspondingly the Lagrange's equation. Using the Rayleigh dissipation function, which can be written for the single generalized coordinate as

$$\Phi = \frac{1}{n+1} c \, |\dot{x}|^{n+1}, \quad c \in \mathbb{R}_+, \quad n \in \mathbb{N}_0, \tag{3.10}$$

allows us to extend the Lagrange's equation (3.2) about the nonconservative kinetic potential $\partial \Phi / \partial \dot{x}$. This will favorably expand (3.2) to

$$\frac{d}{dt} \frac{\partial L}{\partial \dot{x}} + \frac{\partial \Phi}{\partial \dot{x}} - \frac{\partial L}{\partial x} = U. \tag{3.11}$$

Note that in the Rayleigh dissipation function (3.10), the c-term constitutes a positive constant coefficient (or function) and n is the exponent (i.e. power) of the resisting (correspondingly dissipative) force. Further we note that while $n \geq 0$ is generally required for a homogeneous dissipation function of degree $(n+1)$, we will purposefully focus on $n \in \{0,1\}$ only. Then, $n = 0$ is corresponding to the Coulomb friction and $n = 1$ to the dissipative forces of the Rayleigh type, i.e. linear in velocity. The latter is conventionally known as a viscous force. For the considered motion system, from Figure 3.1 (b), one can evaluate (3.11) with $L = T - V$. For $c := \sigma$ and $n = 1$ one obtains (3.8), while for $c := C_f$ and $n = 0$ the resulting dynamics is

$$m\ddot{x} + C_f \operatorname{sign}(\dot{x}) + kx = 0. \tag{3.12}$$

Worth noting is that the Rayleigh dissipation function can also be composed as superposition of both, i.e. $\Phi = \Phi_0 + \Phi_1$ with Φ_0 assuming $n = 0$ and Φ_1 assuming $n = 1$. This will lead to the well known combination of the Coulomb and viscous friction terms, cf. (2.7), this way resulting in

$$m\ddot{x} + \sigma\dot{x} + C_f \operatorname{sign}(\dot{x}) + kx = 0. \tag{3.13}$$

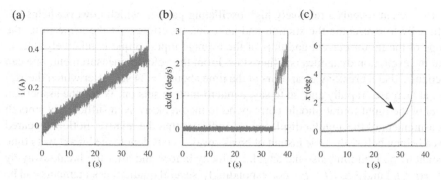

Figure 3.4 Illustrative example of the start of relative motion subject to a gradually increasing input drive torque: measured motor current (proportional to the produced drive torque) in (a), angular velocity in (b), and angular displacement in (c).

3.2 MOTION ONSET AND MOTION STOP

In the previous section, we succeeded to include the kinetic friction into the general form of Lagrangian systems and, based on that, to derive the dynamic equation for an example of the damped mechanical oscillator, with viscous-type friction, see (3.8), Coulomb-type friction, see (3.12), and a linear combination of both, see (3.13). In all those cases, one can solve the initial value problem, i.e. one can find an unique solution of $x(t), \dot{x}(t)$ trajectories for each pair of the given initial values $x(0), \dot{x}(0)$. It is noteworthy, however, that for (3.12) and (3.13) this is not as trivial due to the discontinuous sign term. This issue we are going to address explicitly in the next following section. But by now, we already learned that the kinetic friction is a superposition of both terms, see (2.7), so that a practical question of an onset and stop of the relative motion can be already posed. We also like to stress that this is quite independent of the particular equations which describe the behavior of a system with friction.

In control engineering practice, it is well known that before a visible, so to say, macroscopic motion begins, a drive actuator has to produce a certain amount of the (static) force that will be sufficient to overcome the so-called *stiction* in a mechanical assembly. Only then, a relative motion will start and that rather abruptly, while this so-called *break-away* phase is usually not straightforward, neither in terms of the dynamic transients nor in terms of the amount of input force required. This fact has obviously to do with both, uncertain Coulomb friction and transient frictional by-effects, cf. with section 2.3. For better understanding what happens with relative motion of a drive body, when it starts from an idle state and is subject to friction, let us have a look at the experimental observations exemplary shown in Figure 3.4. A gradually increasing input drive torque, which is proportional to the measured motor current shown in the plot (a), is applied to a motor drive with one unconstrained rotational degree of freedom. The induced angular motion is measured by a precise absolute encoder, as shown in the plot (c). The angular velocity, shown in the plot (b), is obtained as a discrete time derivative of the encoder signal. The measured motor current reveals a relatively high oscillating pattern, which however helps additionally to overcome the stiction and adhesion by-effects. At the same time, the slope of the motor current and thus of the average input torque is relatively flat, so that the excitation dynamics is rather slow. From the velocity measurement, one can recognize that a break-away appears at the time about $t = 34$ sec, after which the $\dot{x}(t)$ signal increases rapidly and, so to say, a macro-motion sets on. While the associated level of the input torque should correspond to the break-away, a sufficiently smooth measurement of the angular displacement indicates that the relative motion appeared already long before the time instant of break-away. Let us denote the break-away time instant by t_b. Denoting also the angular velocity before and after the break-away by $\bar{\dot{x}} \equiv \dot{x}(t < t_b)$ and $\hat{\dot{x}} \equiv \dot{x}(t > t_b)$, correspondingly, several qualitative statements can be made. First, an exact time instant t_b cannot be set or fixed by some threshold value, and it is rather characterized by a relationship $|\bar{\dot{x}}| \ll |\hat{\dot{x}}|$. While an average amplitude of $\bar{\dot{x}}$ is close but not equal to zero, a zero macroscopic acceleration can be assumed before break-away, i.e. $\bar{\ddot{x}} \approx 0$. Furthermore, with a look on Figure 3.4 (c) one can conclude that $\dot{x}(t)$ was continuously increasing in amplitude. It becomes also obvious that $\dot{x}(t) \to 0$ as $t \to 0$ and $\dot{x}(t) \to \hat{\dot{x}}$ as $t \to t_b$, cf. Figure 3.4 (c). Denoting further the dynamic state of relative displacement, for the time $t < t_b$ between the system is at rest and break-away, by $z(t) \equiv x(t) - x(0)$, the total kinetic friction can be assumed as a function of both the velocity and relative displacement, i.e. $F(\dot{x}, z)$. For the considered input ramp, with some constant parameter g describing the input slope and the above assumptions for $t < t_b$, one can approximate the overall motion dynamics by

$$F(\dot{x}, z) = gt. \tag{3.14}$$

We should notice that this is valid only for a quasi-static system behavior before the break-away, i.e. within a relatively small range of displacement which is often referred to as *pre-sliding range* or *pre-sliding regime* of the kinetic friction. Differentiating, with respect to the time, the left and right hand sides of (3.14) and taking into account a nearly zero acceleration in average, one obtains

$$\frac{\partial F}{\partial z} \dot{x} \approx g. \tag{3.15}$$

Obviously, for a constant input slope, i.e. $g = $ const, the $\partial F / \partial z$ term should grow towards infinity for $\dot{x} \to 0$ and, correspondingly, the $\partial F / \partial z$ term should decrease towards some small constant for $\dot{x} \to \hat{\dot{x}}(t_b)$, cf. Figure 3.4. This apparently transient friction behavior, which is predominant before break-away of the relative motion, can easily be understood by means of the force-displacement diagram shown in Figure 3.5. For both, the initiation and reversals of a relative motion, the smooth force transition curves depend on the pre-sliding state variable z which is reset every time the direction of motion changes. This way, $z(t)$ describes an instantaneous distance (understood as pre-sliding distance) to the last position where the sign of $\dot{x}$ changed. Here, we should not stick to any particular modeling approach for describing $F(z, t)$, even though several pre-sliding friction models are available in the literature. It is much more important to fix the following properties which are common for describing (z, F) pre-sliding transition curves.

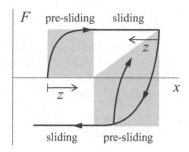

Figure 3.5 Force-displacement map of pre-sliding friction with reversals.

- In the pre-sliding regime, the tangential friction force depends rather on the relative displacement than the velocity.
- Each motion reversal produces a new hysteresis branch in the force-displacement coordinates.
- A pre-sliding distance z is the main factor characterizing the friction force at the beginning or reversal of relative motion.

In the following, we should keep in mind the discussed above pre-sliding friction mechanisms, while the most basic assumptions we made about the kinetic friction in section 2.3 remain valid and dominating at the same time.

Another essential aspect of pre-sliding, which is certainly related to our discussion, is how an ongoing relative motion will stop when it is not additionally supplied by some external (control) input. This question is inherently close to the problem of convergence of a controlled motion, more specifically positioning, once the nonlinear friction is in place. While we will deal with this problem in more detail later, in section 3.3 and chapter 4, it is still relevant here to elaborate qualitatively on the motion stop of a free autonomous system without control. Such consideration should also help understand the transitions from continuous motion to an idle state with $\dot{x} = 0$. Let us first specify three principally differing cases of the pre-sliding friction behavior. Note that this is less related to physical phenomena as such, but much more to the model assumptions that can be made to describe it.

1. In the first case, that we call the Coulomb one and label by C-letter, the friction force changes discontinuously as the direction of motion changes. This behavior is driven by a friction force value which is proportional to the sign of relative velocity, cf. with (2.4).
2. In the second case, we will assume some linear stiffness of the contacts within pre-sliding. Therefore, the friction force value will be proportional to z until saturating at the Coulomb friction level C_f. A relatively high stiffness coefficient Γ, which leads to $F = \Gamma z$ in pre-sliding, is associated with an average stiffness of the junctions of both frictional surfaces. Such piecewise linear and saturated force behavior can be directly mapped by the stop-type Prandtl-Ishlinskii operator, and we label it by PIO-letters.

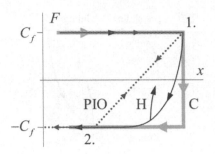

Figure 3.6 Force-displacement behavior of the C-, PIO-, and H-cases of pre-sliding friction.

3. In the third case, we will claim the most generic and also realistic scenario of $F(z)$ pre-sliding friction curves, as addressed above and shown in Figure 3.5. That type of the modeled force-displacement characteristics is determined by hysteresis, and we label it by H-letter. It is also worth noting that this case inherits some properties from the first two cases. The contact stiffness at the boundaries of pre-sliding is $\Gamma \to \infty$ as $z \to 0$, and $\Gamma \to 0$ as $|F| \to C_f$. Moreover, the Coulomb friction law applies as $F \to C_f \operatorname{sign}(\dot{x})$ once system is leaving the pre-sliding range.

The corresponding force-displacement trajectories at motion reversal are schematically shown in Figure 3.6 for the three cases specified above. One can recognize that if a force transition curve of the H-case becomes more stiff, it will proceed closer to the C-case. On the contrary, if an instantaneous stiffness of the H-case becomes lower, it will behave closer to the PIO-case. A natural conclusion is obvious that both cases C and PIO are the limiting cases of H, which is the most general.

An understanding of how all three above cases affect the modeled dynamics of the system during pre-sliding can be further enhanced by analyzing the mechanical work W produced by the frictional force between two reversal points. Recall that the mechanical work

$$W = \int_X F(x)dx$$

will account for the energy dissipated along the path X due to the friction force $F(x)$. Let us consider two arbitrary reversal points labeled by 1. and 2. in Figure 3.6. This way, a motion trajectory 1. $\to$ 2. $\to$ 1. constitutes a closed path $X \equiv \{x_1, x_2, x_1\}$ that coincides with pre-sliding range of the PIO-case. Evaluating first the mechanical work of C-case, where $F(\cdot)$ does not explicitly depend on x but on $\operatorname{sign}(\dot{x})$, results in

$$W_C = -C_f \int_{x_1}^{x_2} dx + C_f \int_{x_2}^{x_1} dx = 2C_f(x_1 - x_2). \tag{3.16}$$

The above expression states unambiguously that the energy is always dissipated by the Coulomb friction. Indeed, even if $(x_1 - x_2) \to 0^+$, that means two consecutive reversals will lie infinitesimally close to each other, the amount of dissipated energy (3.16) remains positive. Due to this fact, the pre-sliding range itself disappears for the C-case. Moreover, it confirms the well-known fact that each switching of a relay-type discontinuous Coulomb friction is dissipative. Now, let us continue this line of argumentation for the PIO-case. Evaluating the energy dissipated by the piecewise linear $F(x) = \Gamma z$ characteristics, cf. with Figure 3.6, one obtains

$$W_{PIO} = \int_{x_1}^{x_2} \Gamma z dz + \int_{x_2}^{x_1} \Gamma z dz = \frac{1}{2}\Gamma z^2 \Big|_{x_1}^{x_2} - \frac{1}{2}\Gamma z^2 \Big|_{x_1}^{x_2} = 0. \qquad (3.17)$$

This indicates that the work produced by $F(z)$ on the way 1. $\rightarrow$ 2. has the same magnitude but opposite sign as the work produced on the way back 2. $\rightarrow$ 1. Apparently, this is due to elastic frictional contacts with the linear stiffness. That means neither dissipation nor supply of the energy occurs alongside the closed X path. This recognition reveals the POI-case as non-dissipative during pre-sliding. It has also a direct consequence for the convergence of free motion trajectories, correspondingly for the motion stop, that is in focus of our discussion. Finally, when evaluating the mechanical work produced by $F(x) = F(z)$ in the H-case, one obtains

$$W_H = \int_{x_1}^{x_2} F(z)dz + \int_{x_2}^{x_1} F(z)dz = \int_{x_2}^{x_1} \big(F(z_2) - F(z_1)\big)dx = \Delta > 0. \qquad (3.18)$$

Here we need to recall that the $F(z_2)$ and $F(z_1)$ force-displacement curves are always differing from each other for two reversal points x_2 and x_1 due to the hysteresis, cf. Figure 3.6. Thus, the evaluated integral (3.18) will result in a non-zero area Δ between both curves. Such area is directly associated with pre-sliding hysteresis loops. Moreover, it reflects the amount of energy losses due to the hysteresis in pre-sliding. At that point, it is worth recalling that C and PIO are both the limiting cases of the H one. If H-case will approach C-case, then $\Delta \rightarrow 2C_f(x_1 - x_2)$. At the other end, if H-case will approach POI-case, the hysteresis loop will collapse, meaning $F(z_2) \rightarrow F(z_1)$ and $\Delta \rightarrow 0$, respectively, leading to $F(z)$ transitions are lossless.

Having gained understanding of the damping properties of all three modeling cases introduced above, we can now look at the associated convergence of the unforced trajectories which are exemplary shown in Figure 3.7. Starting from one and the same non-zero initial velocity, the trajectories behave differently first after the first motion reversal, meaning once the sign of $\dot{x}$ changes. The trajectory of the C-case stays on the x-axis forever once having reached it. There is namely no energy recuperation here within pre-sliding range. The associated finite-time convergence to the invariant set $\dot{x} = 0$ is the same as for the generic differential equation $\ddot{y} + \text{sign}(\dot{y}) = 0$. The PIO-case indicates clearly an undamped harmonic oscillator once the trajectory lands into pre-sliding region. This is not surprising either, since we saw the corresponding energy balance between two reversal points, cf. (3.17) and Figure 3.6. Here, the underlying generic differential equation is $\ddot{y} + y = 0$. The H-case provides a hysteresis-damped convergence of the state trajectory after the first motion reversal,

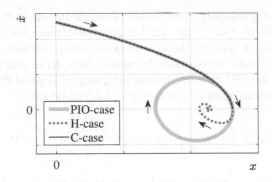

Figure 3.7 Unforced motion trajectories for C-, PIO-, and H-cases of pre-sliding friction.

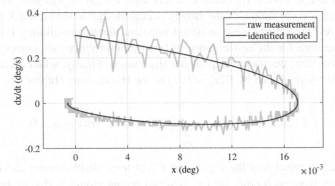

Figure 3.8 Experimental example of precisely measured displacement-velocity phase portrait in vicinity to the motion stop without control, compared with the model fit.

while few transient oscillations can appear depending on a particular $F(z)$ mapping. Recall that the H-case is a more generic and physically faithful one, for which an illustrative experimental example, see [34], shown in Figure 3.8 offers a compelling and practice-oriented interpretation.

3.3   INVARIANT SET DUE TO COULOMB FRICTION

As we have seen before, even if only qualitatively for C-case in Figure 3.7, an unforced trajectory affected by the Coulomb friction remains in an idle state once zero velocity is reached. The final state position $x(t_f)$, where t_f indicates the time instant when $\dot{x}(t)$ becomes zero, depends on the initial values $x(0), \dot{x}(0)$ and can be everywhere on the x-axis. Therefore, an invariant set $\dot{x} = 0$ is natural for an autonomous drive system. Indeed, any real motion will stop somewhere if only an initial velocity but not a driving (control) force is given. More interesting from a control point of view, however, is the question of where a feedback-controlled motion will stop in case of the Coulomb friction. One can also ask whether an exact convergence to zero equilibrium, like we saw it in Figure 3.3 for the system (3.8), can be guaranteed if the latter is extended by the Coulomb friction.

Let us now assume that a drive system with discontinuous Coulomb friction is state-feedback controlled, with the corresponding control gains K_1 and K_2. Worth recalling is that for second-order systems this situation corresponds to a standard PD (proportional-derivative) output feedback controller. If we neglect the reference signal and consider only stabilization of the output position, which corresponds to a setpoint reference when redefining (correspondingly shifting) zero position, the closed-loop dynamics writes

$$m\ddot{x} + K_2\dot{x} + C_f\text{sign}(\dot{x}) + K_1x = 0. \qquad (3.19)$$

We know that through a dedicated assignment of the control gains, for instance by the pole placement, the linear part of the system dynamics (3.19) can be shaped as critically damped, i.e. (3.19) will have two negative real poles at the same location if $C_f = 0$. Thus, it is easy and straightforward to achieve the required convergence of the controlled output position $x(t)$ without Coulomb friction, and that also without transient overshoots or oscillations. Yet, since $C_f \neq 0$ is in place, we need to understand what happens with the system dynamics (3.19) when reaching the x-axis outside the desired equilibrium point $x = 0$. Since the x-axis represents a discontinuity manifold $S = \{x \in \mathbb{R} \setminus 0, \dot{x} = 0\}$, i.e. the states where the sign-operator of the Coulomb friction provides discontinuity in (3.19), one can look at the vector-field of $\dot{x}$ and $\ddot{x}$ dynamics when approaching S from both sides, i.e. once from the half-plane with $\dot{x} > 0$, and once from the half plane with $\dot{x} < 0$. Rewriting (3.19) in a state-space form and evaluating the corresponding vector-field h for $\dot{x} = 0$ one obtains

$$h^+(S) = \lim_{(x,\dot{x}) \to S, \dot{x} > 0} \begin{pmatrix} 0 \\ -K_1x - C_f \end{pmatrix}, \qquad (3.20)$$

$$h^-(S) = \lim_{(x,\dot{x}) \to S, \dot{x} < 0} \begin{pmatrix} 0 \\ -K_1x + C_f \end{pmatrix}. \qquad (3.21)$$

From both vector-field equations given above, it can be recognized that the velocity vectors are pointing in opposite directions for $|x| \leq C_f/K_1$, as schematically drawn in Figure 3.9. Since both vector-fields are normal to the manifold S, neither a continuous motion nor sliding-mode will occur within this interval. Correspondingly, any state

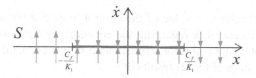

Figure 3.9 Vector-field at discontinuity manifold inside and outside of the largest invariant set, as a result of the Coulomb friction force.

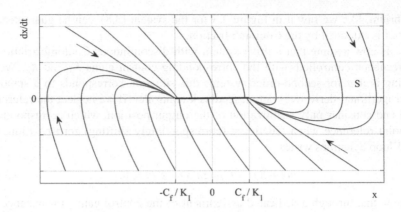

Figure 3.10   State trajectories of the system (3.19) for different initial values.

trajectory will remain sticking for all times $t \geq t_f$, once reaching zero velocity within the interval

$$-C_f/K_1 \leq x \leq C_f/K_1.$$

One needs to emphasize that the above interval constitutes the largest invariant set due to the Coulomb friction, and this coincides with the position error-band of a state-feedback control system (3.19). It can also be noted that this error-band can be reduced by increasing the control gain K_1. At the same time, it cannot be steered to zero as this would require an infinite control gain, that is practically impossible. At the same time, from (3.20) and (3.21) one can recognize that both vector-fields are pointing in the same direction for $|x| > C_f/K_1$. This causes trajectories to cross zero velocity and progress continuously when outside the largest invariant set given above. An illustrative phase portrait of the system (3.19) is exemplary shown in Figure 3.10, cf. with [27].

As next, we want also to qualitatively study the convergence of a drive system with Coulomb friction when it is feedback controlled by a standard PID (proportional-integral-derivative) regulator, with the corresponding control gains K_1, K_3, and K_2, cf. with the PD regulator in (3.19). Considering the drive system plant (2.3), with the kinetic friction as it was analyzed in section 3.2, the resulting dynamics of the closed-loop control system is given by

$$m\ddot{x} + K_2\dot{x} + \sigma\dot{x} + F(\dot{x},z) + K_1 x + K_3 \int x dt = K_1 p + K_2 \dot{p} + K_3 \int p dt. \qquad (3.22)$$

Here $p(t)$ is the set reference value of the control loop. For the sake of simplicity, and without loss of generality regarding the convergence behavior of (3.22), let us consider a set-point position control problem with $p = 0$. Then, differentiation of the left and right hand sides of (3.22) results in

$$m\dddot{x} + (K_2 + \sigma)\ddot{x} + \underbrace{\frac{\partial F}{\partial \dot{x}}\ddot{x} + \frac{\partial F}{\partial z}\dot{x}}_{\frac{d}{dt}F(\dot{x},z)} + K_1\dot{x} + K_3 x = 0. \qquad (3.23)$$

In accord with the analysis provided in section 3.2, we can largely neglect the first partial derivative of F in (3.23), mainly due to $\partial F/\partial \dot{x} \approx 0$ for nonlinear Coulomb friction, and reduce (3.23) to

$$m\dddot{x} + (K_2 + \sigma)\ddot{x} + \left(K_1 + \frac{\partial F}{\partial z}\right)\dot{x} + K_3 x = 0. \qquad (3.24)$$

Obviously, once the closed-loop system (3.24) is in sliding, that implies $\partial F/\partial z = 0$ in accord with section 3.2, its dynamic response can be arbitrary shaped by a dedicated selection of the K_1, K_2, K_3 control gains. This is not surprising and is consistent with the fact that a PID control can effectively compensate for kinetic friction during a steady-state motion. The situation is completely different when the system (3.24) is in pre-sliding, and $\partial F/\partial z$ not only varies largely but can even grow to infinity in the reversal instants. In such situations, the convergence of (3.24) will be largely impaired. This can be assessed qualitatively by, for example, evaluating the corresponding poles of the associated characteristic equation

$$ms^3 + (K_2 + \sigma)s^2 + \left(K_1 + \frac{\partial F}{\partial z}\right)s + K_3 = 0.$$

It is easy to verify that if $\partial F/\partial z \to \infty$, one of the otherwise well-placed poles moves toward zero, so the convergence time-constant also grows toward infinity.

Even if assuming the discontinuous Coulomb friction, i.e. $F(\dot{x}, z) = C_f \, \mathrm{sign}(\dot{x})$, similar as it was done for the PD feedback control in (3.19), one can show that the system (3.22) becomes always sticking when

$$\left| K_3 \int x dt + K_1 x \right| \leq C_f \qquad (3.25)$$

is satisfied. Hence the inequality (3.25) describes the state subspace of a frictional stiction, cf. with the largest invariant set analyzed above for the PD feedback control. Unlike the situation with PD controller, the PID regulated system (3.22) will not remain within this stiction region for always, since the continuously growing integral action will serve for (3.25) becomes (temporary) violated, and the controlled motion is resumed. However, this results in a sequence of the stick-slip cycles, with the $|x|$ value becoming smaller than before with each subsequent period, and the integral control component requires a longer time for (3.25) becomes violated again. Illustrative numerical examples of such stick-slip convergence are shown in Figure 3.11. Alternating or equal sign of the controlled output at convergence depends on the relationship between all control parameters and Coulomb friction coefficient. At the same time, the value of the Coulomb friction coefficient itself does not significantly change the principal stick-slip behavior, cf. Figure 3.11 above. Also the convergence with equal sign of the controlled output implies the long-term stick-slip behavior, as it is visible on the logarithmic scale in Figure 3.11 below.

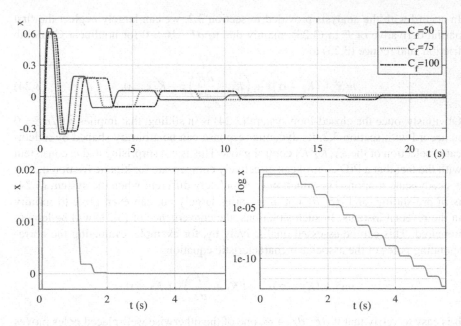

Figure 3.11 Stick-slip convergence of the PID controlled drive system (3.22) for $p = 0$, with alternating sign of x and various Coulomb friction coefficients (above), and equal sign of x once on the linear and once on the logarithmic scale (below).

4 Compensation of Friction in Motion Control

Now, in a sequel to what we have been considering so far, we will deal with compensation of kinetic friction in the controlled motion systems. If we do so, we should be aware that this application relevant topic has attracted attention in the broad academic and practical engineering communities for already more than three decades. Still, kinetic friction, as one of the most cumbersome and less straightforward dynamic disturbance effects, requires continuous investigation and refinement of the appropriate compensation methods. We will start with classical linear feedback controllers which include an integral control action and can, this way, guarantee for a complete cancelation of the Coulomb and viscous friction during the steady-state motion only. Then, we will switch from a problem of tracking the motion trajectory to a more challenging, with respect to the friction compensation, task of the set-point position control. In such control problem, the nonlinear friction with stiction must be overcome, while one needs to ensure the controlled system converges sufficiently fast to the set reference value. Two rather principally simple compensation strategies will be discussed here, one based on the output relay feedback, and another based on the observation of dynamic friction state and its matched feed-forwarding.

4.1 INTEGRAL FEEDBACK CONTROL FOR STEADY-STATE MOTION

The simplest and, at the same time, most common strategy to compensate for frictional disturbances during a steady-state motion is to use an integral feedback term in the velocity control loop. Considering separately the Coulomb- and viscous-type friction, since both are acting in parallel, cf. (2.7), once can first write the linear plant of the drive as

$$m\ddot{x} + (\sigma \pm \delta)\dot{x} = u, \tag{4.1}$$

where an uncertainty of the viscous friction is captured by $0 \leq \delta < \sigma$. The corresponding force to velocity transfer function is then given by

$$G_v(s) = \frac{sx(s)}{u(s)} = \frac{(\sigma \pm \delta)^{-1}}{m(\sigma \pm \delta)^{-1}s + 1}, \tag{4.2}$$

from which one can see that the stationary system gain, i.e. $|G_v(j\omega)|$ for $\omega \to 0$, depends directly on the uncertain viscous friction coefficient $(\sigma \pm \delta) > 0$. This situation is exemplary shown in Figure 4.1. Applying a standard PI (proportional-integral) velocity feedback control, with two tunable control gains $K_p, K_i > 0$, the closed-loop dynamics results in

$$m\ddot{x} + (K_p + \sigma \pm \delta)\dot{x} + K_i x = K_p r + K_i \int r \, dt, \tag{4.3}$$

DOI: 10.1201/9781003415015-4

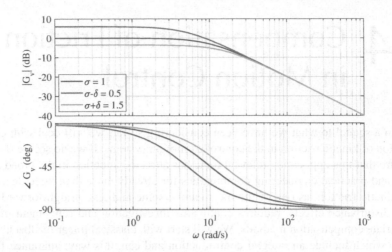

Figure 4.1 Frequency response function of (4.2) for varying viscous friction coefficient.

where the reference velocity value is $r(t)$. For investigating the dynamic control response, and thus also stability of the closed-loop control system (4.3), let us look at the corresponding characteristic polynomial equation

$$ms^2 + (K_p + \sigma \pm \delta)s + K_i = 0. \tag{4.4}$$

We recall that its roots determine the system poles

$$\lambda_{1,2} = \frac{-b \pm \sqrt{b^2 - 4mK_i}}{2m} \quad \text{with} \quad b = K_p + \sigma \pm \delta \tag{4.5}$$

and, therefore, the convergence properties of the velocity output under control. For the physical system parameters and control gains to be always positive, the control system (4.3) can be shown to be always asymptotically stable. This conclusion follows directly from the fact that the inequality

$$\text{Re}\{-b + \sqrt{b^2 - 4mK_i}\} < 0 \tag{4.6}$$

will always hold for the poles (4.5), which are given in their general parameterizable form. This implies that both poles of the closed loop will always lie in the left half of the complex plane, no matter how the positive control gains are assigned. After the stability of the closed-loop system (4.3) is clarified, one can also ask about optimal assignment of the control gains and, furthermore, how the uncertainty of the viscous friction might affect performance of the velocity controller. In order to reply to this question, we first make a pole placement in a way that the closed-loop system (4.3) has two real poles coinciding with each other. This will ensure that there are no transient oscillations and overshoots, on the one hand, and a possibly fast transient response (i.e. since no dominant pole appears) is achieved, on the other hand. Such a

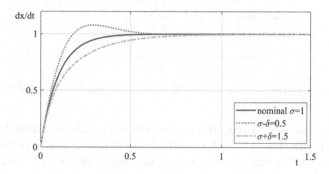

Figure 4.2   Overdamped and underdamped response of the PI controlled velocity, depending on the viscous friction uncertainty factor δ.

pole configuration causes the closed-loop system to be critically damped and is well known when doing a feedback control design. For the imaginary part of both system poles (4.5) to become zero, one requires

$$K_p - 2\sqrt{mK_i} - \sigma \mp \delta. \tag{4.7}$$

At the same time, one can recognize that the K_p parameter determines also the locus of the critically damped pole pair since

$$\mathrm{Re}\{\lambda_{1,2}\} = -\frac{K_p + \sigma \pm \delta}{2m}. \tag{4.8}$$

Obviously, without the uncertainty factor δ, the critically damped pole pair can be placed arbitrarily, by choosing the K_p and K_i control gains with respect to (4.7) and (4.8). Of course, in a practical control design, the limiting factors such as, for instance, actuator constraints and measurement noise must also be taken into account. Since the unknown uncertainty factor δ is in place, one can perform the pole placement and the corresponding assignment of the control gains for a nominal case only, i.e. assuming temporarily $\delta = 0$. Certainly, a persistent $\delta \neq 0$ is then expected to downgrade the optimal transient performance of the nominally designed PI velocity controller. Indeed, the transient behavior of the closed-loop control system becomes either overdamped, i.e. with one slower dominant pole, or underdamped and, therefore, transiently oscillating. This will depend on the sign of uncertainty factor δ during the controlled response. Such nominal and downgraded transient behavior is exemplary shown in Figure 4.2 for the unity step.

Now that we have seen how a PI velocity control deals with uncertain viscous friction, let us also look in detail at the integral feedback attenuation of the Coulomb friction. The nonlinear Coulomb friction, which depends (dynamically) on the direction of relative motion only, reduces the magnitude of the control signal by the same constant amount as long as there is a steady motion without changing direction.

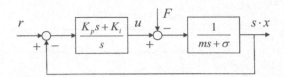

Figure 4.3 Closed control loop of PI velocity control with matched frictional disturbance.

Then, for the closed control loop, this can be seen as a matched constant disturbance, as shown in Figure 4.3. Due to the matched disturbance character of the constant Coulomb friction F, which sign depends on the sign of velocity, one can recognize that

- if $|F| < |u|$, then the control effort will be negatively biased by the constant C_f, thus, resulting in an entirely slower system dynamics;
- if $|F| \geq |u|$, then there is no continuous motion, and one is forced to consider solutions in the Filippov sense, this due to discontinuities in the dynamics when the sign of output becomes alternating.

Considering the first of the above cases, which is characteristic for a continuous steady-state motion, let us take a closer look on the disturbance sensitivity function

$$S(s) = \frac{s \cdot x}{F} = \frac{G_v(s)}{1 + PI(s)G_v(s)} = \frac{s}{ms^2 + (K_p + \sigma)s + K_i}. \tag{4.9}$$

Worth noting is that (4.9) is eligible only as long as the otherwise discontinuous F does not change its sign. Recall that, conventionally, the sensitivity function provides an insight on how the matched disturbance will affect the output value under the feedback control. From $S(j\omega)$, exemplarily shown in Figure 4.4, one can recognize that it is mostly the K_i-gain which affects the Coulomb friction compensation. Indeed, for

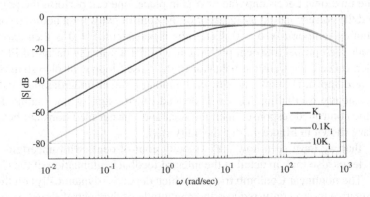

Figure 4.4 Disturbance sensitivity function S for the varying K_i control gain.

$\omega \to 0$, which characterizes the steady-state control system behavior, there is no impact on the output velocity. This implies the constant Coulomb friction disturbance is fully compensated by the integral control term during an unidirectional motion. It is important to accentuate, however, that this is only valid as long as the direction of motion is not changing. At the same time, once the motion direction changes, the closed-loop dynamics becomes excited by disturbance with the step of amplitude $2C_f$. Then, the controlled velocity will converge (again) according to the sensitivity function characteristics (4.9), cf. Figure 4.4.

4.2   OUTPUT RELAY FEEDBACK COMPENSATOR FOR POSITIONING

In section 3.3, we already saw why a single PD feedback controller is unable to guarantee exact convergence of the output position in presence of the nonlinear Coulomb friction. In order to overcome stiction inside of the interval $v = p \pm C_f / K_p$, where p is a constant set reference position, an extension of the output feedback action is supposedly required. Such extension must force the system dynamics sufficiently strong, so as to overpower the Coulomb friction level and, thus, to allow the relative motion to take place also within the v-interval. Intuitively, a sufficiently large relay feedback action, which will depend only on the sign of position control error, can do it in an 'aggressive' but at the same time robust manner. Both, a robustification and aggressiveness of the control action obviously come from a step-like switching. Following to that, the overall position control law can be given by

$$u(t) = K_p e(t) + K_d \dot{e}(t) + K_r \mathrm{sign}\big(e(t)\big), \qquad (4.10)$$

where $e = p - x$ is the position control error. In order to override the Coulomb friction force, the parametric condition $K_r > C_f$ must be fulfilled for the relay control part. It is also worth noting that the control law (4.10) requires the reference value $p(t)$ to be at least once differentiable, i.e. $p(t) \in \mathscr{C}^1$. Also the corresponding feedback gains $K_p, K_d > 0$ must be properly assigned so that the linear part of the closed-loop dynamics, i.e. the one without nonlinear Coulomb friction and relay-based feedback, is asymptotically stable and sufficiently damped. If the first condition placed on the reference value cannot be met, an alternative control formulation

$$u(t) = K_p e(t) - K_d \dot{x}(t) + K_r \mathrm{sign}\big(e(t)\big) \qquad (4.11)$$

for the reference-free derivative part can equally be used.

The control law (4.10), or its alternative (4.11), assumes that the underlying feedback control system without nonlinearities

$$H(s) = \frac{R(s)G_x(s)}{1 + R(s)G_x(s)} \qquad (4.12)$$

is asymptotically stable. Here it is understood that $R(s)$ is an appropriately designed linear feedback regulator (for instance the PD one), and the input-output transfer function of the linear part of the plant is

$$G_x(s) = \frac{x(s)}{u(s)}.$$

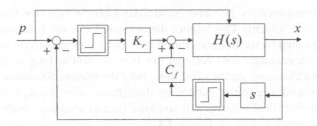

Figure 4.5 Block diagram of the closed-loop control system with the Coulomb friction and relay feedback nonlinearities.

Apparently, for such closed control loops, a subsequent inclusion of both nonlinearities, i.e. Coulomb friction and relay feedback compensator, will lead to a block diagram structure as shown in Figure 4.5. Since the static nonlinearities in feedback, in our case the switching relays, can produce some persistent harmonic oscillations and, thus, obstruct the convergence of the output position, we need to analyze the harmonic balance equation

$$-\frac{1}{\Omega(A_1)} = H(j\omega). \tag{4.13}$$

Here ω and A_1 are the angular frequency and amplitude, respectively, of the first harmonic of the periodically oscillating output $x(t)$. The related describing function Ω depends on the amplitude of the oscillating output, i.e. $A = (\max x(t) - \min x(t))/2$, and is equal to $4a/(\pi A)$ for the relay nonlinearities with relay output gain a. From the harmonic balance analysis, which methods belong to the classical nonlinear control theory, we can recall that if a solution of the harmonic balance equation (4.13) exists, the output $x(t)$ will exhibit the non-vanishing oscillations of the corresponding amplitude A_1 and frequency ω. Graphically, the existence of such solution can easily be verified by inspecting the intersection point of the Nyquist plot of $H(j\omega)$ with $-1/\Omega(A_1)$ plot of the describing function. We also recall that $-1/\Omega(A_1)$ is nothing but the negative reciprocal of the corresponding describing function, as exemplary visualized in Figure 4.6 for Ω that we will address below.

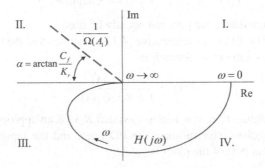

Figure 4.6 Exemplary Nyquist plot of $H(j\omega)$ with $-1/\Omega(A_1)$ plot of the relays feedback.

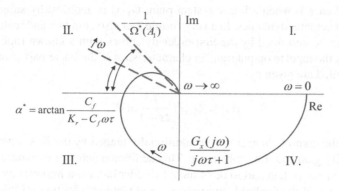

Figure 4.7 Exemplary Nyquist plot of a linear system plant with additional actuator dynamics and $-1/\Omega(A_1)$ plot of the relays feedback.

Coming back to our closed-loop system shown in Figure 4.5, and assuming $p = 0$ for the harmonic balance analysis, one can obtain the overall describing function of both relays in superposition. Because of the ratio $A_2/A_1 = \omega$, which results from Laplace relationship $s \cdot x(s)$ between the output position and relative velocity, the overall describing function of both relays connected in parallel is given by

$$\Omega = \Omega_1 + j\omega\Omega_2 = \frac{4K_r}{\pi A_1} + j\omega \frac{4C_f}{\pi A_2} = \frac{4}{\pi A_1}(K_r + jC_f). \tag{4.14}$$

It is worth noting that graphically $-1/\Omega(A_1)$ represents a straight line that starts at the origin and progresses in a negative direction within the II-nd quadrant of the complex plane, as shown in Figure 4.6. The corresponding angle α and, therefore, slope of the line is equal to $\arctan(C_f/K_r)$. When analyzing the Nyquist plot of $H(j\omega)$, one can recall that for a linear feedback control part of (4.10), (4.11) the second-order plant $G_x(s)$ will result in the also second-order system $H(s)$. Therefore, the Nyquist plot of $H(j\omega)$ will proceed in the III-rd and IV-th quadrants only, and there is no intersection point with $-1/\Omega(A_1)$ graph. Following to that, there exists no (ω, A_1)-solution of (4.13), and the overall closed-loop control system as in Figure 4.5 will converge robustly to zero output error, i.e. $e(t) \to 0$, provided $K_r > C_f$.

With respect to all said above, it is important to emphasize that, however, for the closed-loop systems $H(s)$ with the order higher then two and, as implication, with the phase characteristics which include $\angle H(j\omega) < -180$ deg, the corresponding Nyquist plot will proceed also in the II-nd (or even I-st) quadrant. In this case, the harmonic balance solution of (4.13) can exist, and the closed-loop system will show the remaining steady-state oscillations around the reference point $p = \text{const}$. This is despite the fact that the oscillations amplitude can be negligibly small and the frequency can be sufficiently high, this way both comparable to a background noise.

Let us illustrate in more details the above mentioned situation, where a linear system will have a relative degree higher than two and, hence, the corresponding Nyquist plot also in the II-nd quadrant, as exemplary shown in Figure 4.7. An often encountered case is when a linear system plant $G_x(s)$ is additionally subject to the feedforward actuator dynamics. In a very simple, but also common and realistic case, the latter can be described by the first-order dynamics with a known time constant $\tau > 0$. Then, the input to output transfer characteristics of the linear part of the system to be controlled are given by

$$x(s) = G_x(s)\frac{1}{\tau s + 1}u(s),$$

that means the control channel $u(s)$ is additionally lagged by the first-order element with the unity gain and time constant τ. This additional actuator dynamics not only renders the linear system part to be of the third order but, most importantly, it makes both relays, i.e. of the feedback compensator and Coulomb friction (cf. Figure 4.5), unmatched to each other. Still, using the block diagram algebra, one can bring both relays back to have the same summation point (meaning making them matched) and obtain the modified describing function, cf. with (4.14), as

$$\Omega^* = \Omega_1 + (j\omega\tau + 1)j\omega\Omega_2 = \ldots = \frac{4}{\pi A_1}(K_r - C_f\omega\tau + jC_f). \qquad (4.15)$$

Comparing both right hand sides in (4.14) and (4.15), one can recognize that the real part of the modified describing function Ω^* becomes also frequency dependent. Nonetheless, it is possible to assume ω in $\Omega^*(A_1)$ as a 'frozen' parameter and analyze the negative reciprocal $-1/\Omega^*(A_1)$ in a similar ways as it was done before for $\Omega(A_1)$, cf. with Figure 4.6. One can recognize that the angle

$$\alpha^* = \arctan\frac{C_f}{K_r - C_f\omega\tau} \qquad (4.16)$$

is growing with an increasing angular frequency ω, see Figure 4.7. As consequence, the intersection point and, therefore, the harmonic balance solution will be shifted toward lower amplitudes A_1 and higher frequencies ω of the steady-state oscillations. This comes in favor of the output relay feedback compensator but, at the same time, also carries the risk of possible stationary and not transient harmonic balance solutions already at lower angular frequencies. The first possible intersection point can be found by equating the angular characteristics $\angle[G_x(j\omega)/(j\omega\tau + 1)]$ and the corresponding angle (4.16) of the negative reciprocal of Ω^*.

The illustrative experimental examples of the controlled position response of a closed-loop system (as it is specified in Figure 4.5) are given in Figure 4.8, cf. with [30]. Here a comparison between the compensated (with $K_r = 1.06C_f$) and the non-compensated (with $K_r = 0$) cases is visualized.

4.3   OBSERVER-BASED FEEDFORWARD FRICTION COMPENSATION

The last and quite general friction disturbance compensation strategy that we want to discuss is based on the observation principles of estimating an internal system state that is not available through acquisition with a dedicated hardware. Since the

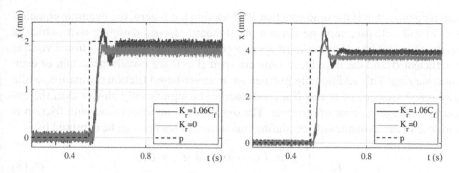

Figure 4.8 Experimental examples of the controlled position response of a drive system with Coulomb friction, for $K_r = 1.06 C_f$ versus $K_r = 0$.

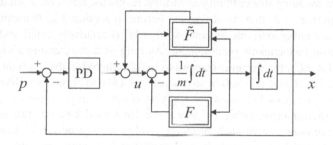

Figure 4.9 Generic control architecture with an observer-based friction compensation.

friction forces and thus the friction state variables are not directly available from measurement, it appears natural to attempt to estimate the current friction state by an observer which should rely on the given knowledge about friction in the system. If one succeeds reconstructing (dynamically) the unknown friction state $\tilde{F}(t)$, then this one can be used for a feedforward friction compensation as schematically shown in Figure 4.9. If a match between the effective and estimated friction values is accurate enough, i.e. $|F(t) - \tilde{F}(t)| \to \varepsilon$, then a standard PD feedback controller will be sufficient for motion control in background. Of course, this is with respect to a particular application at hand which will then determine an acceptable error level $\varepsilon > 0$. It is important to remember that an underlying friction model, that forms basis of the applied friction observer, must be capable of generalization and, at the same time, accurate enough. In the following, we will consider an approach of describing both the linear viscous and nonlinear Coulomb friction as two dynamic state variables that are in superposition to each other, cf. with section 2.3.

For the viscous friction term, one can write

$$\dot{F}_v = \beta^{-1}(\sigma \dot{x} - F_v), \tag{4.17}$$

instead of (2.5), thus capturing the well-known friction lag. Recall that the friction lag appears (also not necessarily as linear) in the relative $(\dot{x}, F_v)$ coordinates. One can recognize that if the time constant β of friction lag is zero, the dynamic equation (4.17) will collapse, and one obtains the standard viscous damping term which is static, cf. with (2.5). Here we must also accentuate that both the proportional viscous coefficient σ and the introduced time constant $\beta > 0$ are usually uncertain or even time-varying. This additionally justifies an observer-based friction estimation, while any time-variances of σ and β are assumed to be significantly slower than the average convergence time of observer. The dynamics of nonlinear Coulomb friction is reduced to the dynamics of pre-sliding transition curves that can be captured by

$$\dot{F}_c = \begin{cases} \dot{x} \cdot \partial f(x,t)/\partial x, & \text{if } |F_c| < C_f, \\ 0, & \text{otherwise.} \end{cases} \tag{4.18}$$

Recall that the modeling of Coulomb friction force, when including also pre-sliding and therefore avoiding discontinuity at motion reversals, relies on a smooth multi-valued mapping $x \mapsto F$ that we discussed before in section 3.2. It means that the relative motion either starts or reverses so that $|\dot{x}|$ is relatively small and the non-viscous friction mechanisms predominate. As long as a progressing $F(x)$-curve is not saturated at $\pm C_f$, the dynamic transitions exist. Otherwise, the constant Coulomb friction force $F_c = C_f \text{sign}(\dot{x})$ will be in place, cf. (4.18) and Figure 3.5. Both dynamic friction states, (4.17), (4.18), allow for applying observation principles, i.e. to use an available prediction error, either $x - \tilde{x}$ or $\dot{x} - \dot{\tilde{x}}$, for a model-based state estimation. Let us first consider one of such observers which reconstructs $\tilde{F}(t)$ based on the difference between the measured $\dot{x}(t)$ and predicted $\dot{\tilde{x}}(t)$ output velocity.

For a general class of the system plants (2.3), with dynamic friction $F(t) \equiv f(\cdot)$ driven by (4.17), (4.18), (2.7), and both available signals $\dot{x}(t)$ and $u(t)$, cf. Figure 4.9, we introduce the state vector $w \equiv (w_1, w_2, w_3)^T := (F_v, F_c, \dot{x})^T$ and derive the state-space model as

$$\dot{w} = \underbrace{\begin{pmatrix} -1/\beta & 0 & \sigma/\beta \\ 0 & 0 & \partial f/\partial x \\ -1/m & -1/m & 0 \end{pmatrix}}_{A} w + \underbrace{\begin{pmatrix} 0 \\ 0 \\ 1/m \end{pmatrix}}_{B} u. \tag{4.19}$$

We should notice that the system matrix is state-dependent, hence one has $A(x)$ in accord with (4.18), while outside the pre-sliding range the $\partial f/\partial x$ term becomes zero. In contrast to that, $\partial f/\partial x \to \infty$ as $x \to x_r$ within pre-sliding range, where x_r is the latest reversal point, cf. with section 3.2. In both borderline cases, the system (4.19) proves to be observable with the output value w_3. Therefore, for obtaining a dynamic estimate of the system states $\tilde{w}(t)$, one can apply the standard Luenberger observer

$$\dot{\tilde{w}} = (A - LC)\tilde{w} + Bu + Lw_3, \tag{4.20}$$

where $C = (0,0,1)$ is the output coupling vector, and $L \in \mathbb{R}^3$ is the vector of observer gains, which are the design parameters. While being excited by both, the control signal $u(t)$ and the plant measurement $w_3(t)$, the state observer (4.20) provides an asymptotic convergence $\tilde{w}(t) \to w(t)$ under one and the single condition. Namely, the

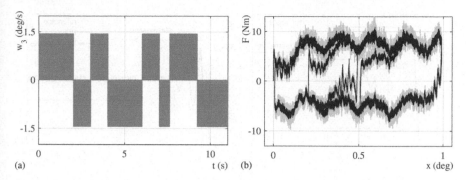

Figure 4.10 Experimental example of the measured observer input $w_3(t)$ in (a), observed friction value $F(t)$ (black line) versus the obtained from the measurement (grey line) in (b).

observer system matrix $(A - LC)$ has to be Hurwitz. Then, the convergence behavior of the observation error vector $e(t) = (w(t) - \tilde{w}(t)) \to 0$ will be solely driven by $\dot{e} = (A - LC)e$, and can be assigned correspondingly, for instance by the pole placement. However, one should not forget about state-dependency of the system matrix A. One can show that designing an observer (4.20) for the case $\partial f/\partial x = 0$ will still lead to $(A - LC)$ remains Hurwitz also for $\partial f/\partial x \gg 0$. Important to notice is that a reverse procedure will not lead to the same result. Therefore, one has first to design the observer (4.20) for the case when the relative motion is outside the pre-sliding range, i.e. for $\partial f/\partial x = 0$. Furthermore, one can also show that the sufficiently damped poles of $(A - LC)$ will, however, migrate and yield one conjugate complex pole-pair when the pre-sliding with $\partial f/\partial x \gg 0$ occurs. Indeed, the dynamics of (4.18), now excited by the observer input $w_3 - \tilde{w}_3$, is not internally damped according to the system matrix $(A - LC)$, cf. with A matrix in (4.19). Through additional injection of an observer damping term D into (4.20), where D is the 3×3 matrix of zeros except $D_{2,2} > 0$ which is the auxiliary design parameter, one can show that the observation error dynamics changes to $\dot{e} = (A - LC)e + D\tilde{w} = (A - LC - D)e + Dw$. Since $w_3(t)$ is the only system measurement entering the observer (4.20) as a measured physical signal, the above Dw-term yields to zero, and one obtains

$$\dot{e} = (A - LC - D)e. \tag{4.21}$$

The resulted observation error dynamics (4.21) is appropriate for both boundaries of the system matrix $A(x)$, i.e. for $\partial f/\partial x = 0$ and $\partial f/\partial x \gg 0$. The design of an D-extended state observer can, therefore, be performed in two consecutive steps. First, one designs L for $\partial f/\partial x = 0$, while taking into account both, the plant dynamics given by A and the effective level of measurement noise of $w_3(t)$. Second, one designs D for $\partial f/\partial x \gg 0$ so that the eigenvalues of $(A - LC - D)$ remain further satisfactory, and that for both boundaries of the system matrix $A(x)$.

An experimental example of the measured observer input $w_3(t)$ and the overall estimated friction force $\tilde{F}(t)$ (black line) versus the one obtained out from the measurement (gray line) are shown in Figure 4.10, cf. [31]. The performance of

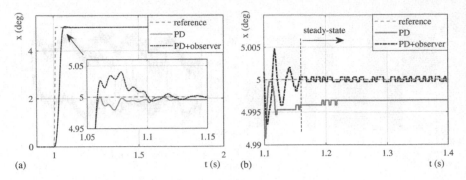

Figure 4.11 Experimental example of the position step response of the control system (Figure 4.9) with and without observer-based friction compensation.

the observer discussed above can be seen from the match between both curves depicted in Figure 4.10 (b). The friction value obtained out from the measurement is derived based on the drive input torque, owing to the quasi-static conditions of a slowly-creeping relative motion, cf. with (2.2). The measured angular velocity $w_3(t)$ discloses a high-frequent binary switching, due to the discrete-time differentiation of the signal obtained from a precise optical encoder. Here it is worth noting that an otherwise low-pass filtering of $w_3(t)$ would represent the actual velocity state. A position step response of the control system as in Figure 4.9 is exemplary shown in Figure 4.11, once with and once without an observer-based friction compensation described above. Zooming in, as shown in plot (b), reveals that the PD controller has a residual non-zero steady-state error. On the contrary, the observer-based friction compensation allows for a fast convergence to zero steady-state error. This is visible from a flipping bit of the optical encoder, see Figure 4.11 (b), the behavior which is corresponding to the sensor resolution.

Despite an already simple structure and effectiveness in application of the friction observer described above, two principal drawbacks may also call for another alternative observation structures. The first drawback is the requirement to measure w_3, which is the relative velocity and often not available for sensing in mechatronic and robotic systems. The second one is the fact that both observation states $\tilde{w}_1$ and $\tilde{w}_2$ are in a superposition to each other, i.e. $\tilde{F} = \tilde{w}_1 + \tilde{w}_2$, cf. with (2.7). This makes their proper decomposition hardly reliable despite a stable convergence of the observer dynamics and a reasonable estimate of the total $\tilde{F}$ value approaching the effective friction force. Moreover, an unavoidable (albeit temporary) occurrence of the conjugate complex poles of the observer (4.20) makes it suboptimal, and can also lead to transient oscillations of the estimated friction states, especially in vicinity of the reversal points. Therefore, we will also discuss another asymptotic observer of motion states with nonlinear friction, which only uses the position measurement and enables a robust design of the estimated state dynamics.

For the same class of 1DOF motion systems (2.3), as considered above, we introduce the state vector $v \equiv (v_1, v_2, v_3)^T := (x, \dot{x}, F)^T$ and derive the simplified state-space model as

$$
\dot{v} = \underbrace{\begin{pmatrix} 0 & 1 & 0 \\ 0 & 0 & -1/m \\ 0 & (\partial F_c/\partial x + \sigma/\beta) & 0 \end{pmatrix}}_{A} v + \underbrace{\begin{pmatrix} 0 \\ 1/m \\ 0 \end{pmatrix}}_{B} u. \tag{4.22}
$$

The associated output coupling vector is $C = (1,0,0)$, since now the relative displacement is the single measured output value, i.e. $x = Cv$. One should notice that
also here, the system matrix A has one varying term $\partial F_c/\partial x$. Thus, we deal again
with $A(x)$ during pre-sliding. Outside the pre-sliding range, the $\partial F_c/\partial x$ term becomes zero. On the contrary, $\partial F_c/\partial x \rightarrow \kappa$ after each reversal point $x(t_r)$, where t_r is
the time instant of the last motion reversal, and $\kappa \gg 0$ is a large positive constant rep-
resenting the initial contact stiffness. Note that one can even have $\kappa \rightarrow \infty$ at t_r. This is
consistent with transient stiction during the motion reversals. For both of the above
cases, i.e. for $\partial F_c/\partial x = 0$ and $\partial F_c/\partial x \gg 0$, the (A,C)-pair proves to be observable in
the Kalman sense. This allows again to design a standard Luenberger-type observer,
for which an asymptotic convergence, i.e. $\tilde{v}(t) \rightarrow v(t)$, can be guaranteed.

In order to simplify the observer design and, moreover, to improve the convergence properties of $\tilde{v}(t)$, we transform the state-space representation (4.22) into the
regular form

$$
\begin{pmatrix} \dot{\bar{v}} \\ \dot{y} \end{pmatrix} = \begin{pmatrix} A_{11} & A_{12} \\ A_{21} & A_{22} \end{pmatrix} \begin{pmatrix} \bar{v} \\ y \end{pmatrix} + \begin{pmatrix} B_{\bar{v}} \\ B_y \end{pmatrix} u, \tag{4.23}
$$

where $\bar{v} = v_1$ and $y = (v_2, v_3)^T$. Then, we are able to apply a standard reduced-
order asymptotic observer. Recall that the above transformation of (4.22) into the
regular form (4.23), with breakdown of A and B into the matrices of an appropriate
dimension, provides a separation into the measurable and unmeasurable states $\bar{v}$ and
y, respectively. Following to that, an exclusion of $\bar{v}$ from the estimated state vector
reduces the order of observer dynamics from three to two and, this way, the overall
complexity of the observer design. The reduced-order Luenberger observer is then
given by

$$
\dot{\tilde{y}} = (A_{22} - LA_{12})\tilde{y} + (A_{21} - LA_{11} + A_{22}L - LA_{12}L)\bar{v}
$$
$$
+ (B_z - LB_{\bar{v}})u, \tag{4.24}
$$

where $L \equiv (L_1, L_2)^T$ is the vector of observer gains, which are the design parameters.
Here, $\tilde{y}$ is the estimate vector of the unmeasurable system states y. It is worth recalling
that the state observer (4.24) provides an asymptotic convergence $\tilde{y}(t) \rightarrow y(t)$ under
one and the single condition. That is the observer system matrix $(A_{22} - LA_{12})$ must
be Hurwitz. Also recall that in order to transform back the dynamic state $\tilde{y}(t)$, the
backwards transformation

$$
(\tilde{v}_2, \tilde{v}_3)^T = \tilde{y} + L\bar{v}, \tag{4.25}
$$

is required subsequently.

As it is claimed above, the reduced-order observer (4.24) has only two estimated variables, for which the corresponding pole assignment is simpler comparing to the asymptotic observer (4.20) considered before. Still, we have to demonstrate robustness of the observer (4.24) against state-variation of the system matrix $A(x)$. We also recall that avoiding transient oscillations in the $\tilde{F}(t)$ estimate is one of the main motivations of the reduced-order observer (4.24). For analyzing the state-variation of the system matrix A, and respectively A_{22}, one can look at the characteristic polynomial of $(A_{22} - LA_{12})$ which is

$$\left(s + L_1\right)s + \frac{1}{m}\left(\sigma/\beta + \partial F_c/\partial x - L_2\right) = 0. \tag{4.26}$$

Obviously, both eigenvalues of (4.26) can be computed explicitly as

$$\lambda_{1,2} = \frac{1}{2}\left(-L_1 \pm \sqrt{L_1^2 + \frac{4}{m}\left(L_2 - \partial F_c/\partial x - \sigma/\beta\right)}\right). \tag{4.27}$$

Based on that, we impose the robust conditions

$$\text{(a)} \quad L_1 > 0 \quad \text{and} \quad \text{(b)} \quad L_1 > 2\sqrt{\frac{\kappa + \sigma/\beta - L_2}{m}} \tag{4.28}$$

and can, this way, ensure the observer dynamics is (i) asymptotically stable and (ii) has no complex poles, which otherwise would lead to transient oscillations during the observer convergence. Moreover, one can recognize that a sufficiently large L_1 will determine the most left hand side pole and, thus, a faster convergence of the $\tilde{v}_2$ estimate, which is the relative velocity. At the same time, L_2 is expected to be negative in several cases, in order to ensure that (4.28) holds also for $\partial F_c/\partial x = 0$. The selected ratio between the L_1 and L_2 gains, satisfying (4.28), is controlling the distance between both real poles (4.27). Note that this distance becomes maximal for $\partial F_c/\partial x = 0$, and it is decreasing during the pre-sliding as the $\partial F_c/\partial x$ value grows. Taken into account the above, the robust observer (4.24), (4.25) can be designed for all transient and steady-state phases of the nonlinear friction, also taking into account the uncertain system parameters, cf. (4.28).

5 Concluding Remarks

In this text, a compact introduction into analysis and compensation of kinetic friction in 1DOF drives, commonly used in the mechatronic and robotic systems, is provided. Despite a long and rich history of studying the frictional effects in the machines and mechanisms, different aspects of cause and effect of the friction forces remain almost always in focus of the system and control engineering. This is not surprising since the phenomena of friction manifest themselves on the weakly specified contact interfaces of the moving bodies. As a logical consequence, frictional effects usually have a spatially distributed, nonlinear, and often time-varying nature and behavioral appearance. This renders the kinetic friction to a challenging disturbance in the drive systems, although it is often matched with the available and used control forces. In addition to the basic assumptions that we made for modeling the viscous and Coulomb friction forces, we have pursued the simplified point-contact and single DOF approximations in our developments. We considered the friction terms of the system dynamics also from the energy dissipation point of view and discussed how the non-conservative friction forces can be integrated into the standard Lagrangian framework of a dynamic system modeling. In addition to a well-understood frictional behavior at steady-state motion, we paid also a particular attention to the non-trivial transients at the beginning and stop and also reversals of relative motion. From a viewpoint of position control, these challenging to analyze phases of relative motion have been in our special focus. In the last part of the text, we discussed a possible control-oriented mitigation of the disturbing friction effects and highlighted both the output-based feedback and observer-based feedforward friction compensation methods. At large, the present text aims to master the basics of modeling and compensation of kinetic friction and to motivate and prepare interested students and engineers for further appealing challenges of friction phenomena.

DOI: 10.1201/9781003415015-5

References

1. Luis Aguilar, Igor Boiko, Leonid Fridman, and Rafael Iriarte. *Self-oscillations in dynamic systems*. Birkhäuser, 2015.

2. Farid Al-Bender and Jan Swevers. Characterization of friction force dynamics. *IEEE Control Systems Magazine*, 28(6):64–81, 2008.

3. Joaquin Alvarez, Iouri Orlov, and Leonardo Acho. An invariance principle for discontinuous dynamic systems with application to a Coulomb friction oscillator. *Journal of Dynamic Systems, Measurement, and Control*, 122(4):687–690, 2000.

4. Brian Armstrong-Helouvry. *Control of machines with friction*. Springer, 1991.

5. Brian Armstrong-Hélouvry, Pierre Dupont, and Carlos Canudas-de Wit. A survey of models, analysis tools and compensation methods for the control of machines with friction. *Automatica*, 30(7):1083–1138, 1994.

6. Derek Atherton. *Nonlinear Control Engineering–Describing Function Analysis and Design*. Workingam Beks, 1975.

7. Ruud Beerens, Andrea Bisoffi, Luca Zaccarian, Maurice Heemels, Henk Nijmeijer, and Nathan van de Wouw. Reset integral control for improved settling of PID-based motion systems with friction. *Automatica*, 107:483–492, 2019.

8. Ugo Besson. Historical scientific models and theories as resources for learning and teaching: the case of friction. *Science & Education*, 22(5):1001–1042, 2013.

9. Robert W Carpick and Miquel Salmeron. Scratching the surface: fundamental investigations of tribology with atomic force microscopy. *Chemical Reviews*, 97(4):1163–1194, 1997.

10. Philip Dahl. Solid friction damping of mechanical vibrations. *AIAA Journal*, 14(12):1675–1682, 1976.

11. Charles-Augustin De Coulomb. *Theorie des machines simples, en ayant egard au frottement de leurs parties et a la roideur des cordages*. Bachelier, nouvelle edition, 1821.

12. Aleksei Filippov. *Differential equations with discontinuous right-hand sides*. Dordrecht: Kluwer Academic Publishers, 1988.

13. Gene Franklin, David Powel, and Abbas Emami-Naeini. *Feedback control of dynamic systems*. Pearson, 8th edition, 2019.

14. Bo Jacobson. The Stribeck memorial lecture. *Tribology International*, 36(11):781–789, 2003.

15. Vincent Lampaert, Farid Al-Bender, and Jan Swevers. Experimental characterization of dry friction at low velocities on a developed tribometer setup for macroscopic measurements. *Tribology Letters*, 16(1):95–105, 2004.

16. Gennady Leonov, Nikolay Kuznetsov, Maria Kiseleva, and Ruslan Mokaev. Global problems for differential inclusions. Kalman and Vyshnegradskii problems and Chua circuits. *Differential Equations*, 53(13):1671–1702, 2017.

17. David Luenberger. An introduction to observers. *IEEE Transactions on Automatic Control*, 16(6):596–602, 1971.

18. Anatoliy Lurie. *Analytical mechanics*. Springer, 2013.

19. Richard Murray, Zexiang Li, and Shankar Sastry. *A mathematical introduction to robotic manipulation*. Taylor & Francis, 1994.

20. Henrik Olsson, Karl Johan Åström, Carlos Canudas-de Wit, Magnus Gäfvert, and Pablo Lischinsky. Friction models and friction compensation. *European Journal of Control*, 4(3):176–195, 1998.

21. Wengen Ouyang, Shivaprakash N Ramakrishna, Antonella Rossi, Michael Urbakh, Nicholas D Spencer, and Andrea Arcifa. Load and velocity dependence of friction mediated by dynamics of interfacial contacts. *Physical Review Letters*, 123(11):116102, 2019.

22. Bo Persson. *Sliding friction: Physical principles and applications*. Springer, 2013.

23. Elena Popova and Valentin Popov. The research works of Coulomb and Amontons and generalized laws of friction. *Friction*, 3(2):183–190, 2015.

24. Ernest Rabinowicz. *Friction and wear of materials*. Wiley, 1995.

25. James Rice and Andy Ruina. Stability of steady frictional slipping. *Journal of Applied Mechanics*, 50(2):343–349, 1983.

26. Michael Ruderman. On break-away forces in actuated motion systems with nonlinear friction. *Mechatronics*, 44:1–5, 2017.

27. Michael Ruderman. Stick-slip and convergence of feedback-controlled systems with Coulomb friction. *Asian Journal of Control*, 24(6):2877–2887, 2022.

28. Michael Ruderman. Energy dissipation and hysteresis cycles in pre-sliding transients of kinetic friction. ArXiv, DOI: 10.48550/arXiv.2212.05799, 2023.

29. Michael Ruderman. Robust asymptotic observer of motion states with nonlinear friction. In *22nd IFAC World Congress*, page 6468–6473, 2023.

30. Michael Ruderman and Leonid Fridman. Analysis of relay-based feedback compensation of Coulomb friction. In *IEEE 16th International Workshop on Variable Structure Systems (VSS)*, pages 95–100, 2022.

31. Michael Ruderman and Makoto Iwasaki. Observer of nonlinear friction dynamics for motion control. *IEEE Transactions on Industrial Electronics*, 62(9):5941–5949, 2015.

32. Michael Ruderman and Makoto Iwasaki. Analysis of linear feedback position control in presence of presliding friction. *IEEJ Journal of Industry Applications*, 5(2):61–68, 2016.

33. Michael Ruderman, Makoto Iwasaki, and Wen-Hua Chen. Motion-control techniques of today and tomorrow: a review and discussion of the challenges of controlled motion. *IEEE Industrial Electronics Magazine*, 14(1):41–55, 2020.

34. Michael Ruderman and Dmitrii Rachinskii. Use of Prandtl-Ishlinskii hysteresis operators for Coulomb friction modeling with presliding. In *Journal of Physics: Conference Series*, volume 811, page 012013, 2017.

35. Andy Ruina. Slip instability and state variable friction laws. *Journal of Geophysical Research: Solid Earth*, 88(B12):10359–10370, 1983.

36. Michael Urbakh, Joseph Klafter, Delphine Gourdon, and Jacob Israelachvili. The nonlinear nature of friction. *Nature*, 430(6999):525–528, 2004.

37. Antonis I Vakis, Vladislav A Yastrebov, Julien Scheibert, Lucia Nicola, Daniele Dini, Clotilde Minfray, Andreas Almqvist, Marco Paggi, Seunghwan Lee, Georges Limbert, et al. Modeling and simulation in tribology across scales: An overview. *Tribology International*, 125:169–199, 2018.

38. Andrea Vanossi, Nicola Manini, Michael Urbakh, Stefano Zapperi, and Erio Tosatti. Colloquium: Modeling friction: From nanoscale to mesoscale. *Reviews of Modern Physics*, 85(2):529, 2013.

39. Victor Zhuravlev. On the history of the dry friction law. *Mechanics of Solids*, 48(4):364–369, 2013.

Index